CB Radio

Construction Proje

by

Len Buckwalter

SECOND EDITION

FIRST PRINTING—1973

International Standard Book Number: 0-672-20952-7
Library of Congress Catalog Card Number: 73-75083

Preface

CB radio was established to provide any citizen with an efficient and low-cost method of business and personal communications. The projects in this book are intended to work along with this idea.

The projects include add-on accessories, operating aids, test instruments to monitor the output of your equipment, and other innovations. The devices are simple and also inexpensive. No prior knowledge of electronics is required to complete these projects, and a minimum of tools is necessary.

Of special value are the instruments which are designed to check the operating strength and overall performance of the transmitted signal. With a 4-watt limitation on the output power of CB transmitters, even small tuning errors or problems in an antenna can severly reduce the output range of the signal. The simple instruments described in this book can be used for regular measurements and checks which are bound to result in consistently better communications.

LEN BUCKWALTER

Contents

CHAPTER 17

CHAPTER 18

1

Construction Techniques

None of the projects described in this book conflict with existing CB regulations which govern what the CB'er may and may not do to his equipment. All aspects of operation are covered in detail in Part 95 of the FCC Rules and Regulations. A copy of the rules should be in the possession of the CB operator.

The principal aim of these rules is to prevent any change in equipment which may result in interference to other stations. The circuit that generally effects interference is the transmitter. The FCC, however, realizing that certain transmitter adjustments may not be completed effectively by the equipment manufacturer, allows the CB operator to adjust his final rf amplifier stage for maximum power output. (A transmitter is most efficient when tuned to a particular antenna at the time of installation.) Customarily, there are one or two tuning screws provided for these adjustments. If the transceiver transmitter complies with Part 95 of the regulations, the adjustments will not result in improper operation.

The most critical section in the transmitter is the crystal-oscillator stage, or frequency-determining circuit. Aside from changing crystals, no adjustments may be made to this stage unless performed by, or under the supervision of, at least a Second-Class Radiotelephone license holder.

The instruction manual provided with most CB units usually indicates the location of the various tuning adjustments. When any of the devices described in this book are put into service, they should be used solely in conjunction with those legal adjustments located in the final rf amplifier stage.

Any adjustments or modifications to the receiver of a CB rig may be performed by the CB'er since these will not result in interference on the band. Also, home-built antennas are acceptable as long as they comply with the height restriction stated in the regulations.

ELECTRONIC PARTS

Components used in the projects are available from electronics parts distributors. Since few of the circuits are critical, some component substitution is practical. Wherever precise values are required, this is noted in the text or parts list. Therefore, if there is a question as to the type of capacitor, the tolerance and wattage of a resistor, or the size of hookup wire to use, it is unimportant (unless noted). For example, a 10,000-ohm resistor that is specified in a parts list can have a tolerance of up to 20% and be rated anywhere up to 1 watt. Capacitors can be paper or ceramic disc. (When not stated, the voltage rating of capacitors should be about double the power-supply voltage. Unmarked ceramic capacitors are 500- or 1000-volt units that may be used in any circuit.)

When the terminals on a component must be identified, consult the illustrations as a guide. The 1N34, for instance, is marked in different ways. (See Fig. 1-1 for a comparison of two systems.) In general, if the meter in a project reads backwards, it is a sign of a reversed diode. Handle diodes with care, especially while soldering. Heat may be conducted away from the diode element by grasping the lead being soldered with pliers or an alligator clip.

In several of the projects, a connector carries energy between the CB transceiver and the assembled device. When the jacks and plugs shown in the photos do not match those

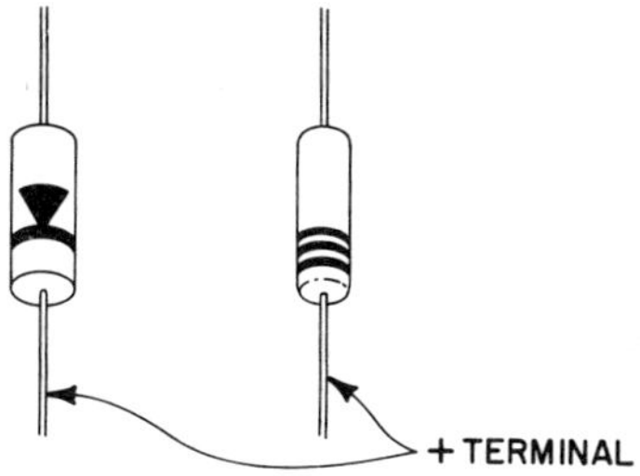

Fig. 1-1. Diode markings.

+ TERMINAL

of your unit, connectors that do match should be substituted. Three typical connectors in use are the coaxial, automobile, and phono types (Fig. 1-2).

The aluminum cases that house most of the projects need not be the exact size indicated. Although the author used standard catalog items, you may choose any adequate substitute you desire. Wood or other cabinets are acceptable *only* if shielding is unimportant. The Output-Power Indicator and Scope Adapter, for example, require metal enclosures to prevent rf energy from radiating and producing possible interference to nearby CB transceivers.

ASSEMBLY

It is important that any given chapter be completely read before beginning construction. In a few cases, a particular

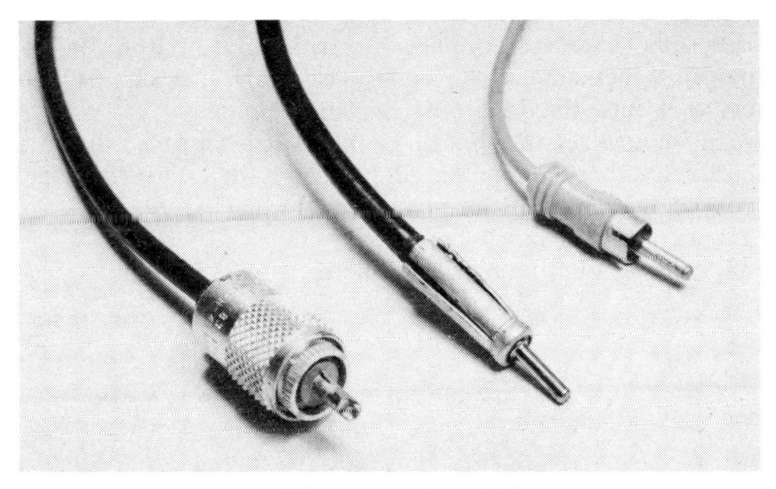

Fig. 1-2. Different types of coaxial fittings.

project will not work with certain CB transceivers, and this information should be known before you attempt assembly. For instance, the devices that connect to a transceiver, such as the S meter, are intended to function with the typical, tube-type CB superhet unit. On the other hand, the test devices that pick up a radiated rf signal, like the field-strength meter, should work with virtually any CB model.

If you are interested in theory, there is a section on circuit operation in each chapter. Even if electronic principles are beyond you this section is still recommended reading since, although devoted mainly to schematic explanations, the discussion on circuit operation may contain helpful construction hints.

TEST EQUIPMENT

Few pieces of test gear are needed for assembly. A multimeter, however, is a handy instrument for checking circuit voltages and for checking wiring errors. Also, when a project contains a tuned circuit that will not resonate in the required frequency range, a grid-dip meter is handy. As an example, differences in wiring layout produce changes that could shift circuit values. In most projects suggestions given in the text are adequate for compensating for these differences through a "trial and error" method. If this method does not produce satisfactory results, however, it is recommended that you use a grid-dip oscillator. By using this instrument you can check the tuned circuit and bring it into the required frequency range.

In one project (the Scope Adapter) it is assumed that the builder will use the finished project with an oscilloscope. Although a scope is rarely owned by a CB operator, the adapter was included because of its ability to equip the scope for presenting a graphic picture of a CB output signal.

2

S Meter

There are at least a half-dozen reasons why a CB receiver should be equipped with an S meter (Fig. 2-1). The primary purpose of this meter is to indicate accurate tuning—to make certain the receiver dial is adjusted for highest reading on an incoming signal. But there are several other important functions possible with an S meter. In a way, it may be considered a built-in test instrument. If you want to discover the effects of a new antenna or try a new mounting location for an old one, the S meter accurately indicates any differences. Simply note the reading from a known station "before and after." Everything else being equal, the S meter reveals any improvement (or deterioration) with far greater precision than is possible by ear. This ability to perceive differences in signal strength is also useful for installations which use a beam antenna. As the antenna is rotated, the meter pinpoints correct antenna orientation for a given station.

Although S meters, including the one about to be described, may be used to report signal strength in "S" units, the method should be considered only comparative. All S meters will not read the same for a given signal. There is an informal standard which states that an S9 reading on the meter means that a 50-microvolt signal is being received. This reference, however, is not usually achieved in CB equipment (it's too costly) and S-meter reports usually acquire the reputation for being "generous" or "stingy."

Nevertheless, the instrument yields helpful readings of a comparative nature that often spot trouble. Let's say that one of your mobile units always causes a reading of S5 when parked at a particular point. If this reading drops, chances are that trouble exists in the mobile transmitting system or your receiving setup. Since the S meter continu-

Fig. 2-1. A homemade S meter.

ously monitors the performance of most of the receiver, it gives lower readings if a defect exists in the antenna or any tube stages, except those in the audio amplifier. In a corresponding manner, trouble in a transmitting section of the distant mobile unit may also cause a drop in your S-meter indications. Thus, the meter should be used to the full extent of its ability to warn of signal changes that point to trouble.

For those who perform their own receiver alignment, the meter is an excellent indicator for peaking the various coils in the rf and i-f sections of the receiver.

CIRCUIT DESCRIPTION

The S meter described here is designed to operate with a superhet CB receiver. As shown in Fig. 2-2, it is a two-transistor device that relies on its own internal battery for power. The extremely light current drain enables the S meter to perform for periods in excess of six months before

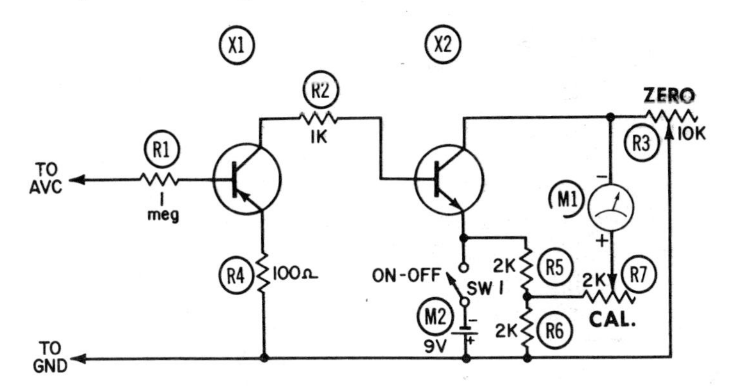

Fig. 2-2. Schematic diagram of the S meter.

the battery needs to be changed. In operation, the circuit picks up the automatic volume control (avc) voltage in the receiver, amplifies it, and then drives the meter. The S-meter circuitry is designed to prevent overloading or shorting of the avc voltage.

Thus, only two connections are required to connect the S meter into the receiver—avc and ground. Due to its independent power source, it can work in any set.

CONSTRUCTION

There are several ways to mount the S meter to the CB set. As shown in Fig. 2-3, the meter and On-Off switch have been mounted in a separate aluminum case which sets on top of the set. The rest of the circuitry is built on a piece of perforated board connected to the meter through a cable.

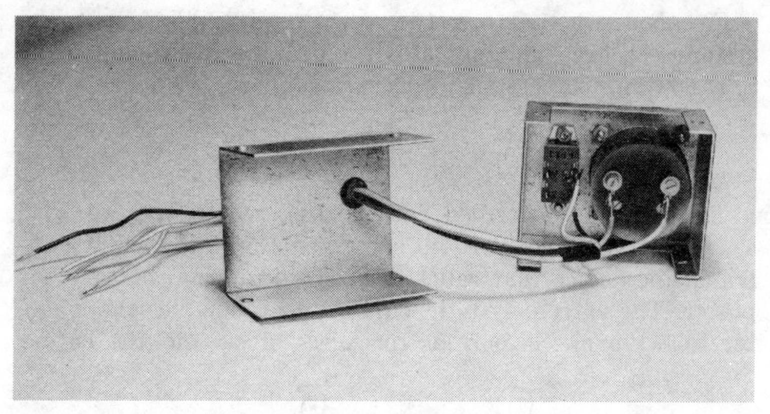

Fig. 2-3. Open view of the S-meter case.

If desired, the complete unit can be housed within the CB cabinet if there is enough room. There are even some cases where the meter can be attached to the front panel after a suitable hole has been cut. In any instance, be sure that the On-Off switch and the Calibrate and Zero controls are easily accessible. These last two might need an occasional readjustment, especially if wide changes in temperature affect the amount of current conducted by the transistors.

The first steps in construction are shown in Fig. 2-4. The small aluminum case has been cut out to receive the On-Off switch and meter. Next, the various wires that make up the

Fig. 2-4. Construction details.

Fig. 2-5. Using a perforated board for mounting components.

interconnecting cable are attached (Fig. 2-3). Follow this system if you decide to construct the unit, as illustrated. Otherwise, make the cable as long or as short as you need for your particular installation.

A convenient method for constructing the two-stage amplifier is on a piece of perforated board (Fig. 2-5). Note that small clips (especially made for this purpose) are in-

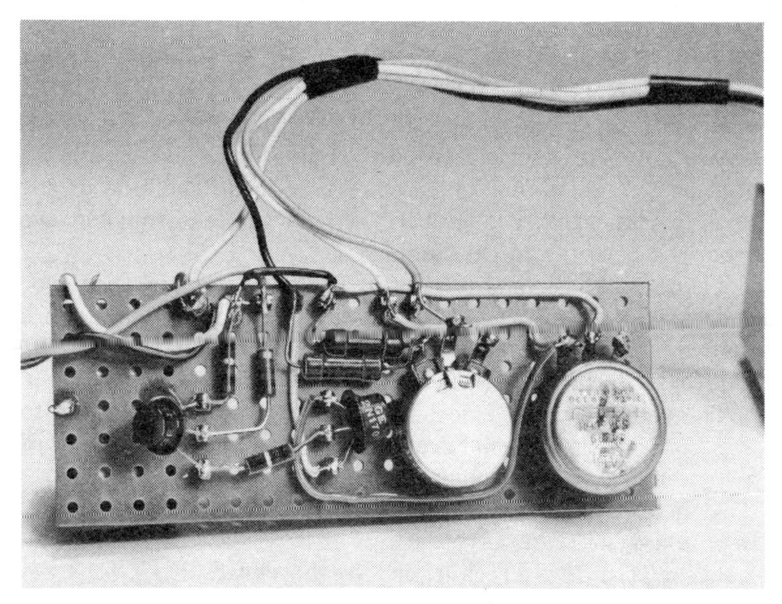

Fig. 2-6. Components in place and wired.

serted into the board at various tie points to support the small components. The completely wired version appears in Figs. 2-6 and 2-7. Notice that the two controls are at the lower right of the board.

The reverse side of the board appears in Fig. 2-8. The two knobs near the left are the calibration (R7) and Zero (R3) controls. The battery is mounted to the right of them by means of a heavy piece of bare wire threaded through holes in the board. The two free leads in the photo connect into the CB set. Be sure to twist them as suggested in the pictorial diagram.

The connections into the CB set are found by means of the schematic in Fig. 2-9 and the schematic diagram of your set. Here is where you can tell if your rig is one of the two or three models that will not work with the S-meter unit. (Some sets impress a high negative voltage on the avc line when operating on transmit. This rules out the S meter since such voltage would cause the meter to conduct excessive current during transmit.) The first step is to locate the last i-f, or intermediate-frequency, transformer on the schematic of your set. Usually, this is the one marked 2nd or 3rd i-f. It immediately precedes the stage marked detector. Once this section is located, compare it with the schematic of Fig. 2-9—a typical avc section used in most receivers. Note that a resistor (valued between 47K and 56K) is tied to one winding of the transformer. This continues down to another resistor that is most often 1 megohm or 2.2 megohms. The avc voltage is on the side of the resistor indicated in Fig. 2-9. Connect the lead marked avc from the S meter to this point.

There will be some variation from one set to the next, but basically the schematic should have the two resistors just described. If they are not present, you have one of the units that is incompatible with the S meter.

The remaining lead goes to ground. Simply connect it to any convenient chassis ground point—a mounting foot of a terminal strip or under a screw head.

Although the S-meter circuit is described for installation in a tube-type transceiver, it might be tried in a transistorized set if a few precautions are observed. First, the set must have a changeover relay (which usually produces an

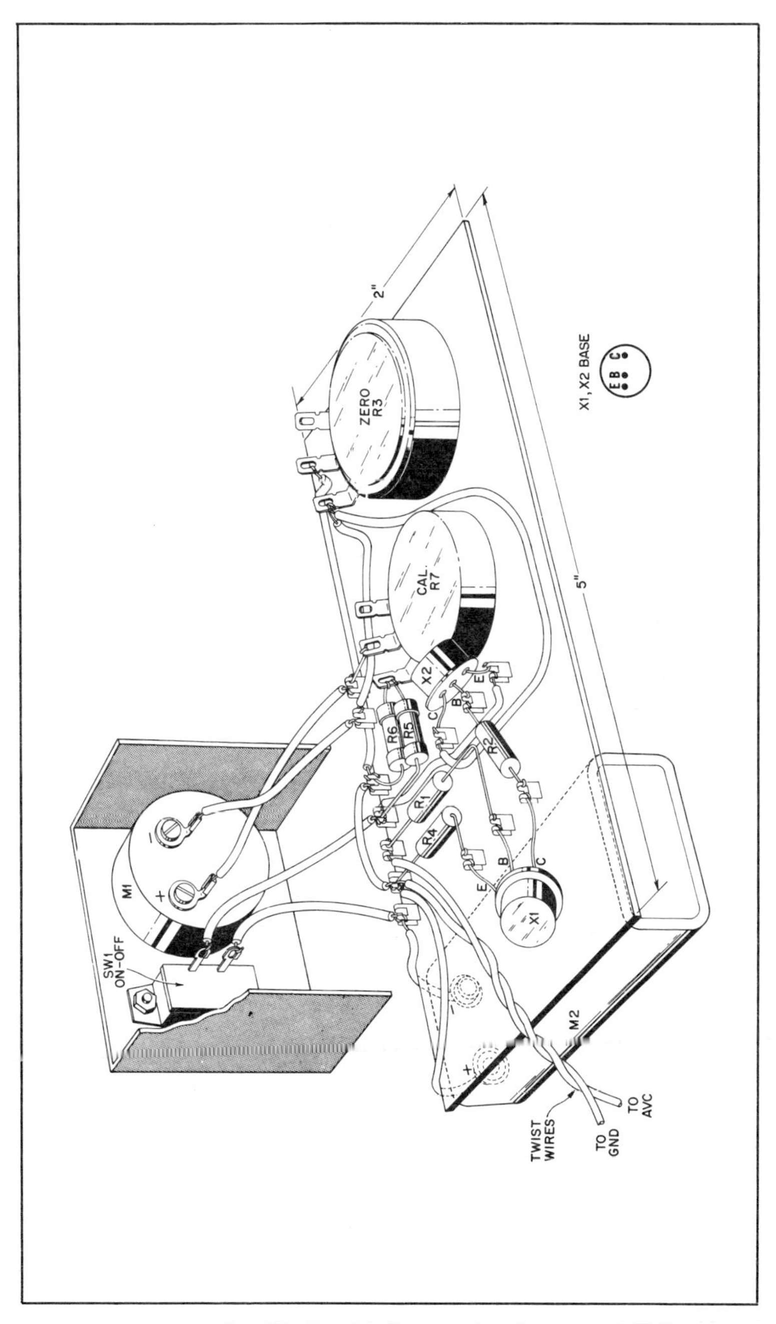

Fig. 2-7. Pictorial diagram of project.

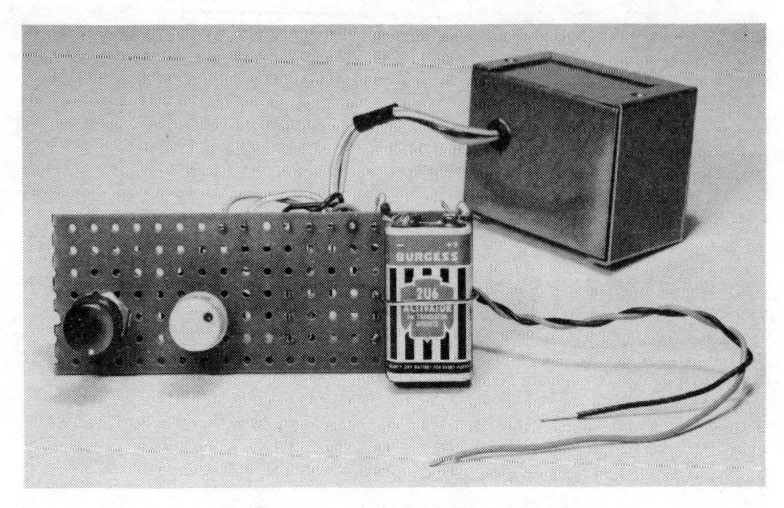

Fig. 2-8. Front view of perforated board.

audible click when the mike button is depressed). This really means that electronic switching, not compatible with the S-meter circuit, isn't being used. Further there must be no contact between the ground (+ side) of the meter circuit and the case or chassis of the transceiver. This is done by avoiding any wiring connections directly from the S-meter circuit to the S-meter case. Finally, if the S meter

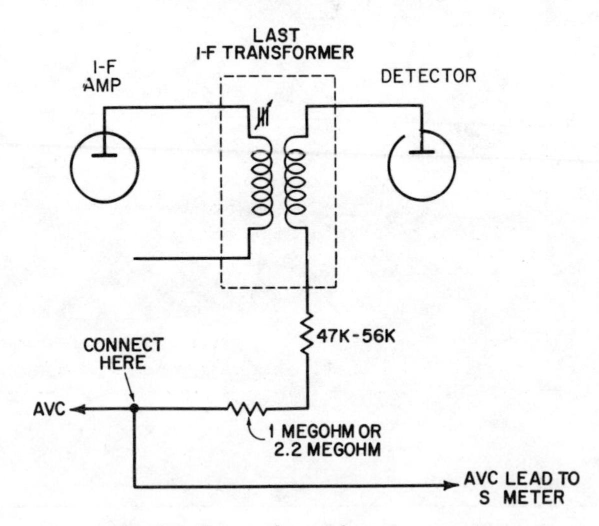

Fig. 2-9. How to locate the avc connection.

reads backwards in operation, reverse the two connections to the CB set.

OPERATION

Once all the interconnections have been made, turn on the CB set and flip the S-meter power switch to On. The meter pin should swing either full up or down. Now disconnect the antenna and begin to adjust the Zero control. It should be possible to bring the pin exactly to the zero mark of the meter. Reconnect the antenna and tune in a CB signal. If it is moderately strong with no atmospheric noise behind it, adjust the Calibrate control to move the meter pin to an S9 reading. (More accurate results are possible when calibration is done against another CB set which has a built-in S meter.) After a period of operation, the proper adjustment should be more apparent. If a station within approximately a quarter of a mile from your location is received, the reading should be nearly full scale, about 30 dB over S9.

PARTS LIST

Item	Description
R1	1 megohm resistor ½ watt.
R2	1K resistor, ½ watt.
R3	10K carbon potentiometer.
R4	100-ohm resistor, ½ watt.
R5, R6	2K resistors, ½ watt.
R7	2K carbon potentiometer.
X1	2N107, GE-2, SK3003, or HEP250 transistor.
X2	2N107, GE5, SK3011, or HEP641 transistor.
M1	0-meter (0-1 mA basic movement).
M2	9-volt battery (Burgess 2U6 or equivalent).
SW1	Spst slide switch.
Misc	Perforated board 5″ × 2″; aluminum case approximately 2¾″ × 1⅝″ × 2″ (optional); two knobs and terminal clips.

3

Portable Antenna

When kept in your car trunk, the portable antenna is ready for any emergency or special event. As shown in Fig. 3-1, it winds up into a small roll that takes up little space when not in service. Then, when you set up your rig, it unfolds into a full-size antenna which can be installed in a matter of minutes. These features make it useful for CB operation in temporary situations—at a campsite or waterfront, for example. There's no need for mounting brackets or other time-consuming installation techniques.

Performance of the antenna is not as efficient as the more elaborate base-station units, but antenna adaptability to special situations more than compensates for any lack of efficiency. Actual use of the antenna has shown that it has approximately a 10-mile range in base-to-mobile operations.

CIRCUIT DESCRIPTION

The antenna operates as a simple half-wave dipole, and its length is selected to fall within the 27-MHz band. Loading coils, radials, or other special matching devices are not required. The two arms of the dipole resonate in the band and present the proper impedance match to the coaxial cable that serves as a transmission line.

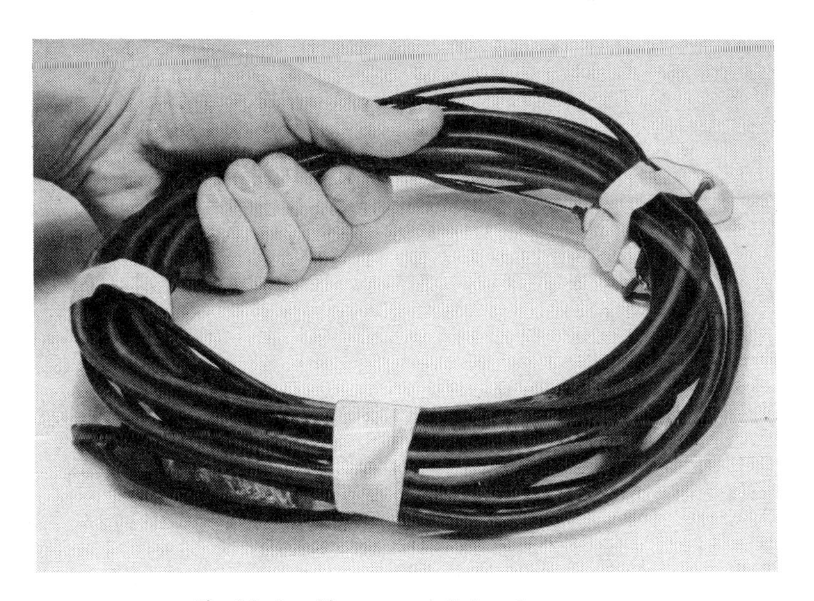

Fig. 3-1. Portable antenna coiled up for storage.

Virtually all CB transceivers are nominally rated at a transmission-line impedance value of 52 ohms. The fact that a 72-ohm coax cable serves as the transmission line in this case has little effect. It was selected to match the antenna rather than the input to the transceiver. The small mismatch from cable to transceiver will, in most cases, not affect range to any significant extent. In fact, most sets are equipped with pi-networks in the transmitter output. Such circuits easily match the 72-ohm coax during the normal tune-up process.

CONSTRUCTION

Assembling the antenna takes about an hour when the specifications given in Fig. 3-2 are followed. All of the components are standard, including one glass and two egg insulators and the coaxial cable (RG/59U). The arms of the dipole, each 104″ (or 8′ 8″), can be ordinary zip cord used for ac wiring. It does not matter if the insulation is left on; just remove a few inches at the ends where they are soldered to the insulators.

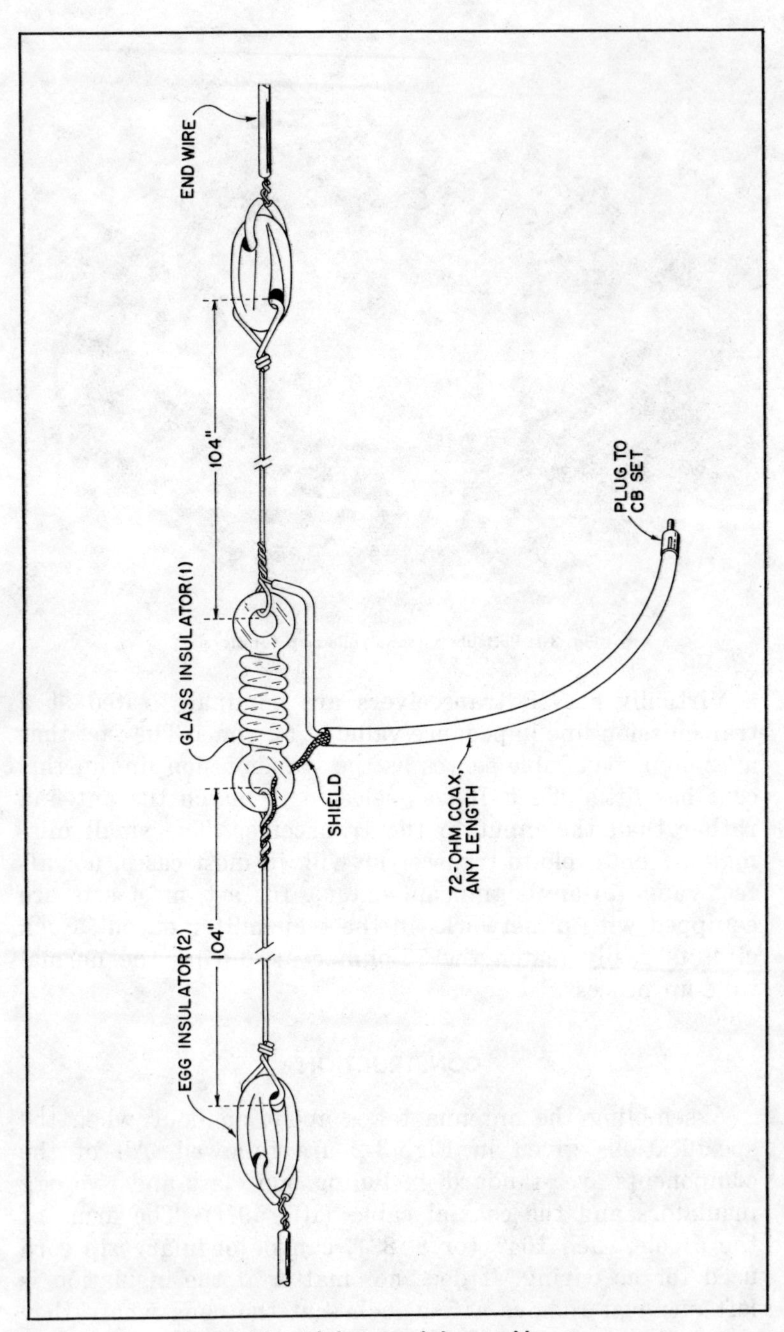

Fig. 3-2. Pictorial diagram of the portable antenna.

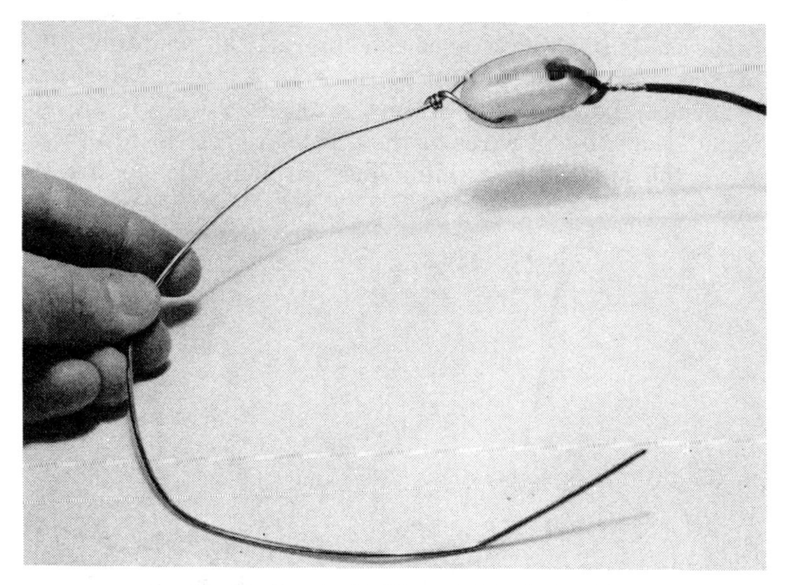

Fig. 3-3. Attaching the "holding wire" to the insulator.

The detail in Fig. 3-2 shows how the coaxial cable connects to the center glass insulator. One arm of the dipole is looped through the eye of the insulator and soldered back on itself. During this operation the shield of the coax is soldered to the dipole arm. The same procedure applies to the other side of the insulator, using the center wire of the coax. The joints at this location will last longer if the shield of the coax is made short enough to bear the weight of the cable rather than the center lead.

Both outside ends of the dipole are treated according to the detail in Fig. 3-3. After the end of the dipole arm is looped and soldered to the insulator, the end wire is soldered to the remaining hole of the insulator in a similar manner. End wires are used solely for mounting the antenna. Preferably cut from solid copper wire (about No. 16 or 18), these end wires are looped and twisted to the supporting points for the antenna.

INSTALLATION

The vertically polarized antenna is standard in CB work. Although dipoles are most often positioned horizontally, best results in this instance call for vertical mounting. One of the simplest mounting sites is in a tree. After two branches, spaced about 17 feet apart (vertically), are selected, the two end wires of the dipole are fastened to them. Also, the side of a building may provide the needed tie points.

Fig. 3-4. Portable antenna in use.

There is an important consideration in running the coax transmission line away from the antenna. Unless it is brought away at right angles, the shield of the coax can interact with the dipole arms and cut down signal strength. The recommended technique is illustrated in Fig. 3-4. (The middle insulator appears near the center of the photo with the dipole arms running above and below it.) Notice how the coaxial cable (at right) is brought away from the antenna at right angles. For proper performance the coax should not bend downward until it is about 8 feet or more away from the antenna.

4

Modulation Monitor

Keeping modulation as close to 100% as possible is an effective method for increasing range. If the CB operator speaks too softly or at too great a distance from the microphone, modulation percentage may drop to inadequate levels. The effect at a remote station may be high S-meter readings, which indicate a strong carrier, but audio that is weak or unintelligible. Excessive audio, on the other hand, may produce overmodulation and interfering "splatter" signals across the band. Intelligibility also suffers as voice tones become harsh and distorted during those periods when the 100% mark is exceeded. The modulation monitor described here (Fig. 4-1) provides a visual indication of each of these undesirable conditions.

The instrument is not intended as a continuous "in-line" monitor. It is, rather, temporarily connected to the output of the CB transceiver (Fig. 4-2) in place of the antenna. Speaking into the microphone causes the needle to jump at a syllabic rate and indicate percentage modulation. This can help to determine the best talking distance and voice level.

CIRCUIT DESCRIPTION

The schematic in Fig. 4-3 shows the internal components of the monitor. The signal from the CB transmitter enters

Fig. 4-1. A homemade modulation monitor.

from the left through coaxial connector, PL1. The rf energy is mostly dissipated across the string of three resistors, R1, R2, and R3. The resistors are a dummy load since the indicator circuit consumes only a tiny amount of power to drive the meter. The rf signal is converted to dc by diode D1.

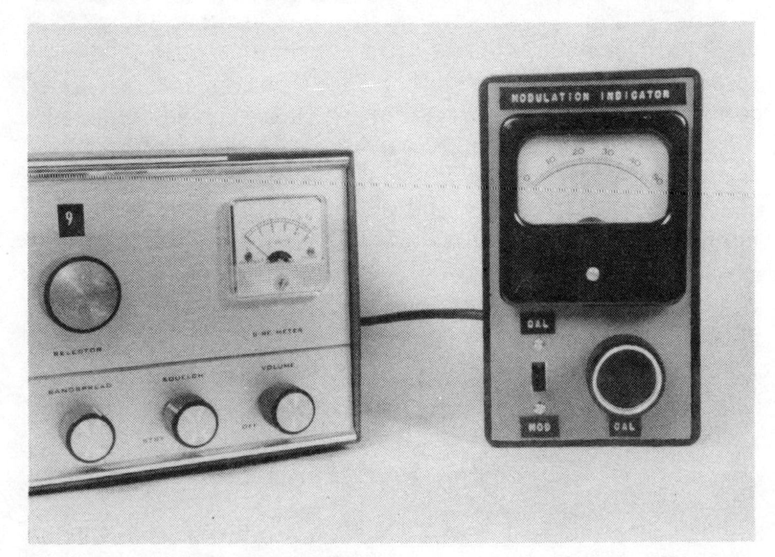

Fig. 4-2. Modulation monitor connected to output of CB transceiver.

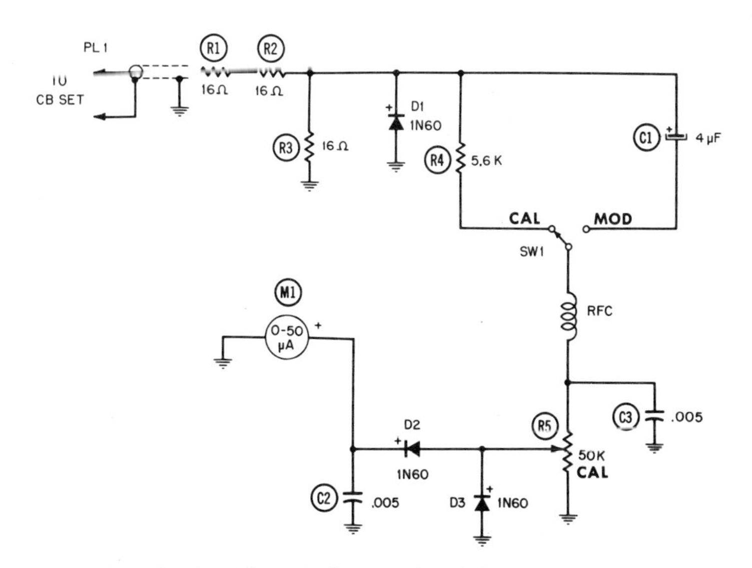

Fig. 4-3. Schematic diagram of modulation monitor.

Assume for a moment that only the steady carrier is present and the operator is not speaking into the microphone. During this time, the instrument is set to its Calibrate position by SW1. As the steady carrier passes through diode D1 it is converted to dc, travels through the radio-frequency choke, and reaches the meter. This provides a reference point for measuring modulation percentage; the calibrating control is turned until the meter reads exactly full scale.

After calibration, SW1 is turned to its Modulation position. Now when the operator speaks into the microphone, the carrier changes strength with modulation. This time, however, the signal must pass through capacitor C1. This blocks the steady carrier and permits only pulsations which represent the audio signal to pass through to the meter. Audio is recovered from the carrier by diodes D2 and D3, and the needle now responds by kicking in step at the voice rate.

CONSTRUCTION

The model shown was built into a bakelite instrument case, but an aluminum case of the same approximate size will serve as well. Mount the various components in suitable holes cut in the front panel, as shown in Fig. 4-4. Since radio frequencies are in the circuit, keep leads short and direct. The pictorial wiring guide in Fig. 4-5 and the actual photo in Fig. 4-6 can serve as a guide to parts and wiring placement.

Observe the correct polarity in wiring the three germanium diodes. The bar end of each diode, marked "+" in the schematic diagram, is identified in various ways on the actual diode itself. There may be a black band or an indentation at the end which corresponds to "+." The electrolytic capacitor must also be installed with regard to polarity. Note how it is marked with positive and negative leads.

A 3-lug terminal strip supports several parts near switch SW1. Be certain the center lug is also the mounting foot for the strip since this also serves as a ground connection if you build the circuit in a metal case. In this model the case is plastic, so the lug merely serves to join several grounds in the circuit, including the shield of the coaxial cable.

Fig. 4-4. Construction details.

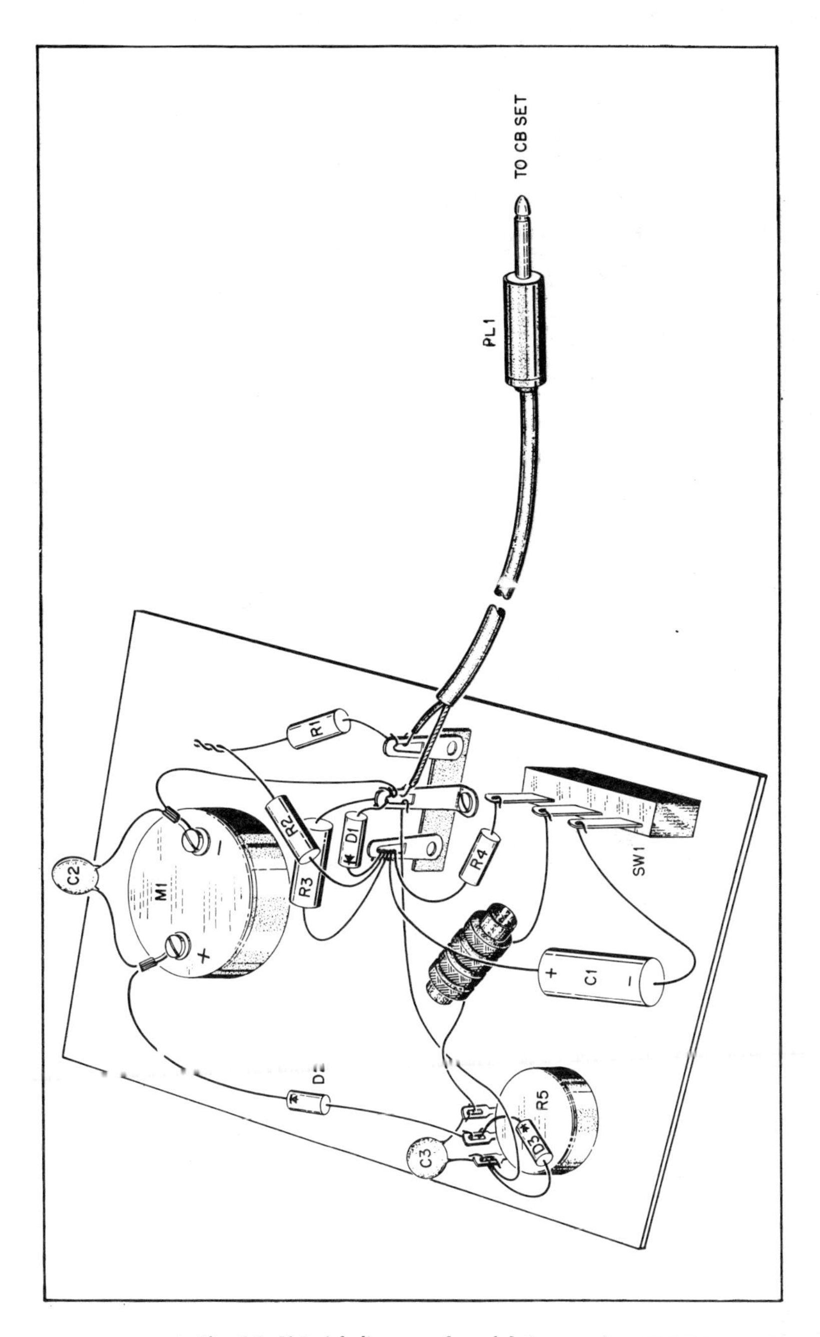

Fig. 4-5. Pictorial diagram of modulation monitor.

Fig. 4-6. Components mounted.

OPERATION

The finished project is checked by connecting it to the antenna socket of a CB transceiver. When the set is placed on transmit, usually by pressing the mike button, there should be a reading on the meter. The Mod-Cal selector (SW1) is placed in the Cal position at this time. Adjust the calibrating potentiometer so the meter needle reads at the top of the scale, or "50."

Next, turn the switch to Mod and speak into the mike. When the needle occasionally hits 35 on the meter, this is

PARTS LIST

Item	Description
R1, R2, R3	16-ohm resistors, 2 watts.
R4	5600-ohm resistor, ½-watt.
R5	50K carbon potentiometer, linear taper.
D1, D2, D3	1N60 germanium diodes.
C1	4-μF electrolytic capacitor, 50 Vdc.
C2, C3	.005-μF disc ceramic capacitor.
PL-1	Coaxial connector (PL-259).
M1	0 to 50-dc microammeter.
RFC	2.5 millihenry choke.
SW1	Spdt slide switch.
Misc	3 feet of 50-ohm coaxial cable; case, 6¼" × 3¾" × 2"; 3-lug terminal strip.

an indication of about 100 % modulation. After using the instrument for a while, you'll be accustomed to the readings when the rig is performing correctly, and note any change that could mean trouble.

5

Multipurpose Test Oscillator

The capabilities of this hand-held instrument (Fig. 5-1) may be applied to a variety of checks in the CB set. Essentially a miniature signal generator, it produces two test signals—a radio-frequency output on 27 MHz and a steady audio tone.

The signals are available separately, depending on the type of test being conducted. The completed model is transistorized and ready for operation at the flick of the power switch.

Possible tests with this unit fall into two basic categories—the receive and transmit sections of the CB rig. A selector switch is placed on RF for receiver checking. In this position, a steady tone-modulated signal is emitted; the frequency depends on the crystal plugged into a socket at one end of the case. (The crystals are the same ones used in the transmit section of the CB unit). With this signal source, overall performance of a receiver may be observed, dial calibration accomplished, or actual alignment carried out.

For the transmitter, the unit provides a source of audio energy on approximately 1000 Hz. This signal, which may

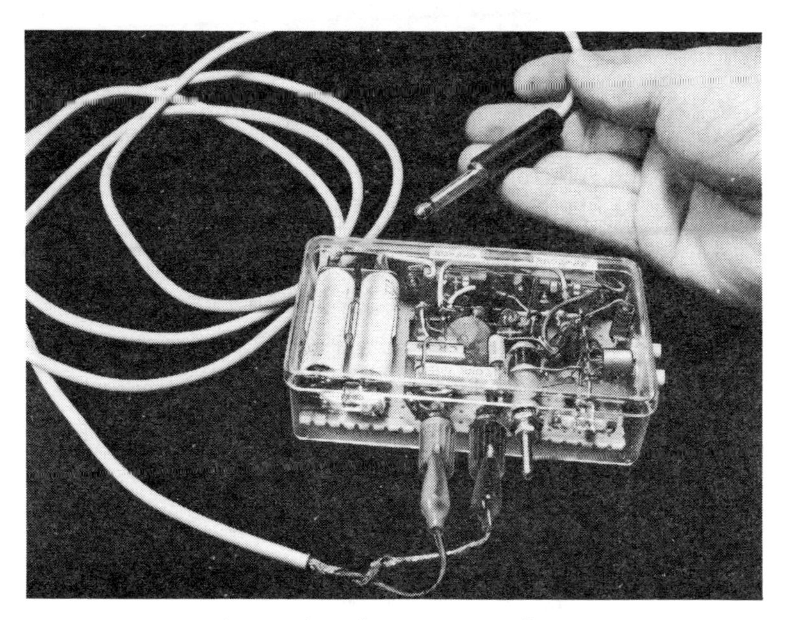

Fig. 5-1. The multipurpose test oscillator.

be fed directly into the mike jack, provides an audio signal for checking modulation more conveniently than speaking into the microphone. The tone is also available for injection across the volume control in the receiver to check for operation of the audio stages. Among the other possibilities for the instrument is checking crystals for oscillation. Any channel crystal plugged in should produce a tone on the proper frequency in the receiver.

CIRCUIT DESCRIPTION

As shown in Fig. 5-2, two transistors function as separate oscillators. Radio frequencies are generated by transistor X1, according to the crystal being used. The other transistor (X2) serves as the audio oscillator for tone generation. When SW1 (the rf-tone switch) is placed in the RF position, both oscillators are activated. Although there is no antenna, adequate rf signal strength radiates from oscillator coil L1 to reach a CB set located within 5 or 6 feet. This explains the use of a plastic case; a metal cabinet would shield against radiation.

The selector switch also brings the tone from the audio oscillator to the collector circuit of transistor X1. Modulation occurs and rf energy is impressed with a tone of approximately 1000 Hz.

When the selector is placed in the Tone position, the rf oscillator is disabled. Audio now appears at binding posts J2 and J3, and tone is available for checking audio circuits in the CB set. Power consumption is extremely low, and the two pen-cell batteries should have extremely long life.

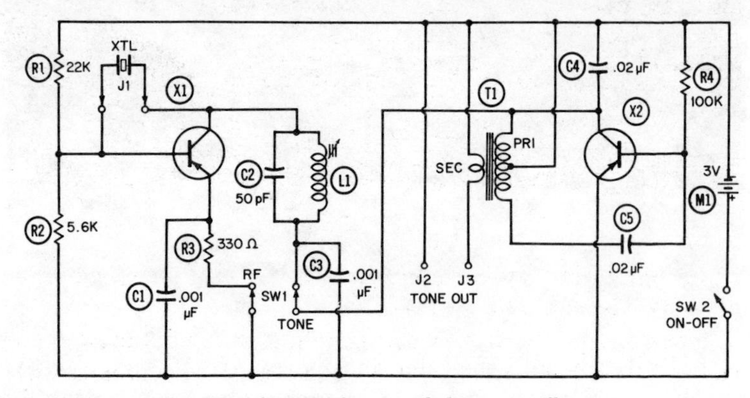

Fig. 5-2. Schematic diagram of the test oscillator.

CONSTRUCTION

Preparing the plastic case for mounting components is not difficult as long as a few precautions are observed. Don't attempt to use a reamer on the plastic to enlarge any openings; this is apt to crack the material, One trick is to use a spinning drill tip as a cutter. After a hole is drilled, it may be easily enlarged by moving the drill bit with a sideways motion until the proper diameter hole is made. As shown in Fig. 5-3, a file is a handy instrument for making straight-sided holes.

For all components to fit properly inside the case, there's a definite sequence for installation. Begin by mounting the two switches, crystal socket, and two binding posts. Once the holes for the binding posts are made, remove the posts and reinstall them when the project is nearly completed. Otherwise, they will not permit the perforated board to slide into position. The hole for the coil is also done at this

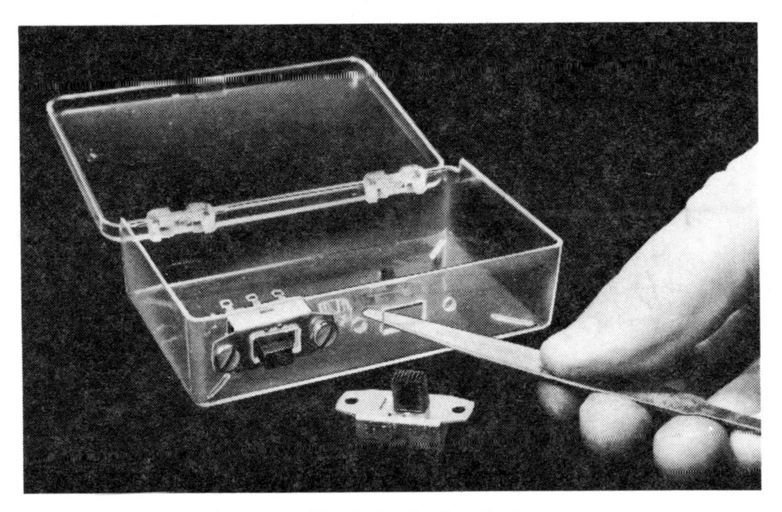

Fig. 5-3. Making holes in the plastic case.

time, but don't leave the coil in place. It, too, is mounted after the board is in position.

Next, the perforated board is held on the plastic case and cut to the inside dimensions. There is no need for a tight fit here; some play is advisable around the edges. Remove the board and add the major components shown in Fig. 5-4;

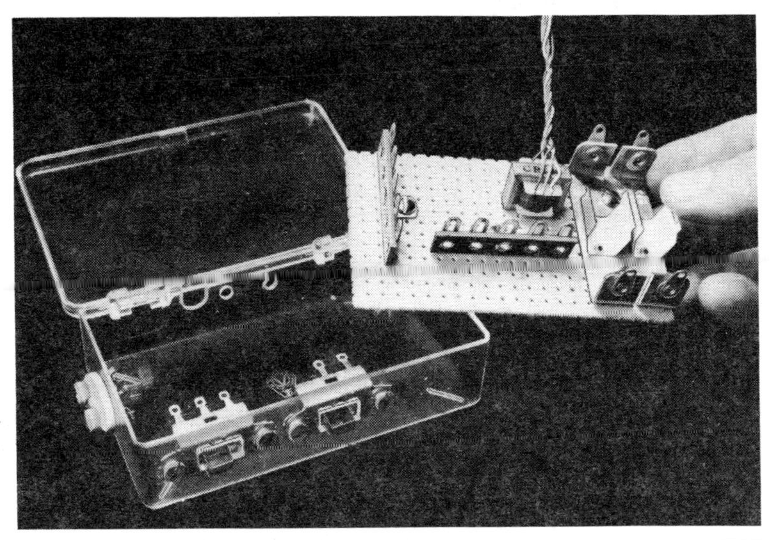

Fig. 5-4. Components mounted on board.

the two terminal strips, audio transformer, and battery holder. After this is completed, the board is permanently positioned in the case. If the lugs on the crystal socket obstruct the board, flatten out these terminals for added clearance. It is not necessary to fasten the board to the case. If solid hookup wire is used, various connections between the board and case-mounted parts hold the board securely in position.

Use the specifications given in the parts list for winding the coil. The job is easy if one end of the enamel wire is first soldered to either coil terminal. (Be sure to scrape away about ¼ inch of enamel before soldering.) Wind the 10 turns close-wound (leaving no spaces between turns). When completed, the coil terminals should be facing in the direction of the crystal socket.

The remainder of the components are soldered to the terminal strips according to the illustrations. Although your audio transformer may have mounting holes, some do use tabs. For the latter type, push the tabs through the holes in the board and bend them flat.

While mounting transistor X1, note that one of its four leads (interlead shield S) is not used. Snip it off close enough to the body of the transistor so there's no danger of it touching adjacent wires.

It's also a good idea not to solder leads to the two binding posts while they are screwed in place—heat from the soldering iron might melt the plastic. To attach these wires, simply unscrew and remove the lugs from their posts and solder them outside the case. A completed view of the project is in Figs. 5-5 and 5-6.

OPERATION

For the first trial with the finished instrument, place the switches in position for generating rf. (The various switch positions should be labeled on the case for future reference and to prevent the possibility of leaving the unit on when not in service.) A CB crystal is inserted in socket J1. Now turn on a nearby CB set and tune to the same channel being used in the instrument. There's a possibility that the tone in the CB speaker will be heard at this time. If it is not,

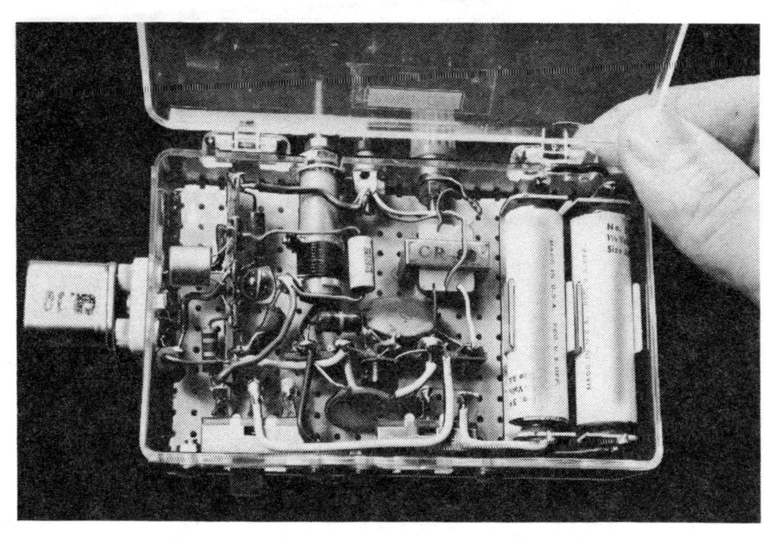

Fig. 5-5. Finished project with CB crystal in socket.

start rotating the tuning screw, or slug, of coil L1. At one point in its travel, the tone should become audible in the speaker. The final adjustment of the coil is best achieved with a receiver equipped with an S meter. The coil is tuned for a peak reading. This setting, however, may not be the most stable. Improved operation results if you back off slightly on the coil tuning until the S meter just begins to drop slightly. Flick the power switch rapidly on and off several times; the tone should return each time.

Due to wiring differences between your model and the original it's possible that your coil will not tune within the proper range. If no tone is heard grasp the two terminals of the coil and twist them very slightly apart. This should cause the coil turns to spread. Don't spread them too far; just enough so a hairline spacing appears between three or four of the coil turns. The tuning process may now be repeated for the proper results.

Another method to determine whether the crystal oscillator is working properly is with a dc milliammeter on a 5- or 10-mA range. Turn off the power switch (SW2) during this hookup since the milliammeter leads will automatically turn the unit on. The positive lead of the meter connects to the switch terminal that goes to the positive

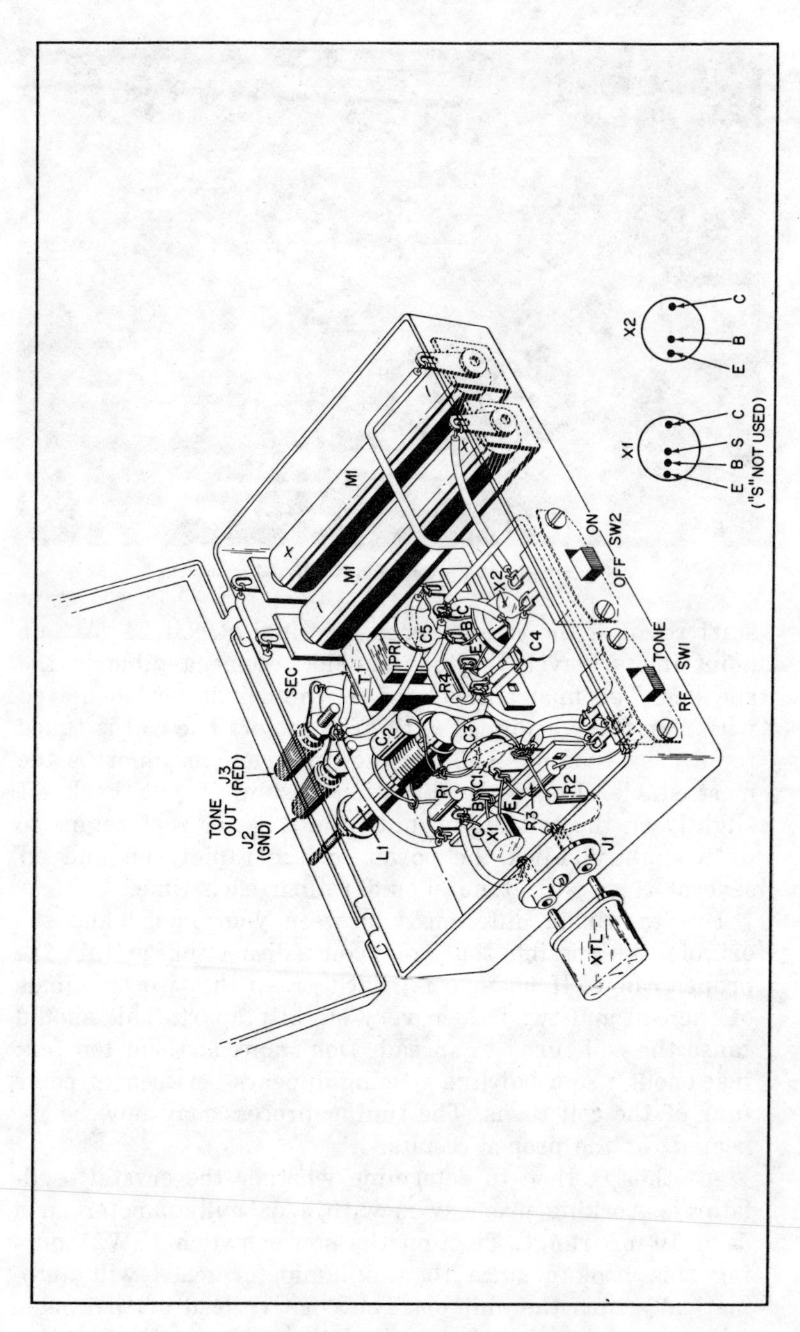

Fig. 5-6. Pictorial diagram of the test oscillator.

battery terminal, the negative meter lead to the other switch terminal. After the meter is hooked up you should note a steady current reading between approximately 1 and 1.5 mA. This indication will suddenly rise by a small but observable amount when the coil slug hits the right setting and the crystal breaks into oscillation. Actually in the original model the slug could be rotated over several turns before the crystal failed to oscillate. The meter check is also a good test of whether the circuit is wired properly. The current reading should be in the vicinity of the value given.

Testing audio output of the device begins by switching SW1 to the Tone position. A pair of leads are secured to the binding posts. The red post (J3) may be considered the "hot" terminal and the black post is ground. If you intend to introduce audio into the microphone jack to check modulation of the transmitter it's best to use a shielded lead.

Audio output as mentioned earlier is useful for testing audio circuits. If, however, you wish to inject tone to a cir-

PARTS LIST

Item	Description
R1	22K resistor, ½ watt.
R2	5.6K resistor, ½ watt.
R3	330-ohm resistor, ½ watt.
R4	100K resistor, ½ watt.
C1, C3	.001-μF capacitor, disc ceramic.
C2	50-pF tubular mica capacitor.
C4, C5	.02 capacitor, disc ceramic.
X1	2N371, GE-9, SK5007, or HEP3 transistor.
X2	2N107, GE-2, SK3003, or HEP250 transistor.
SW1	Switch, slide type, double-pole single-throw.
SW2	Switch, slide type, single-pole single-throw.
J1	Crystal socket with a pin spacing of .486", and a pin diameter of .050" (National type CS-7).
J2, J3	Binding posts, one red, one black.
T1	Audio output transformer, miniature transistor type, 500-ohms center tapped primary to 3.2-ohm secondary.
L1	Coil wound on ⅜" slug-tuned coil form. 10 turns of No. 22 enameled copper wire.
M1	3-volt battery (2 size AA cells in series), with holder.
Misc	Two 5-lug terminal strips; plastic box, 4½" × 2¾" × 1¼"; perforated board to fit.

cuit that might have high dc voltage (the plate of an audio amplifier for example), an additional component is required. This is a 0.1-μF paper capacitor rated at 600 volts. It should be wired in series with the audio transformer lead that runs to the red terminal post (J3).

6

Headset Adapter

The option to choose between speaker and headphone listening is a desirable feature. Earphones allow incoming messages to remain private and prevent speaker sounds from disturbing other people in the home or office. Also, phones can improve intelligibility in mobile operations—where a high noise level from a car or boat may mask speaker sound. The only unit that won't work with the adapter is the type which contains a speaker that doubles as a microphone.

CIRCUIT DESCRIPTION

As shown in Fig. 6-1, the adapter unit is contained in a small metal box. Its purpose is to provide a convenient physical and electrical match between headphone and audio output from the transceiver. Audio from the output transformer in the radio is fed to jack J1 in Fig. 6-2. When the adapter is not in use, the CB circuit is automatically wired for normal operation; audio from the transformer reaches the speaker.

When plug PL1 is inserted into J1, however, audio from the transformer is diverted to the adapter. Applied across volume control R1, sound becomes available for driving the earphone. The action of J1, a closed-circuit jack, causes the

Fig. 6-1. Headset adapter.

speaker to silence whenever the adapter is in service. Removing PL1 immediately restores speaker sound. Potentiometer R1 controls the sound level of the earphone.

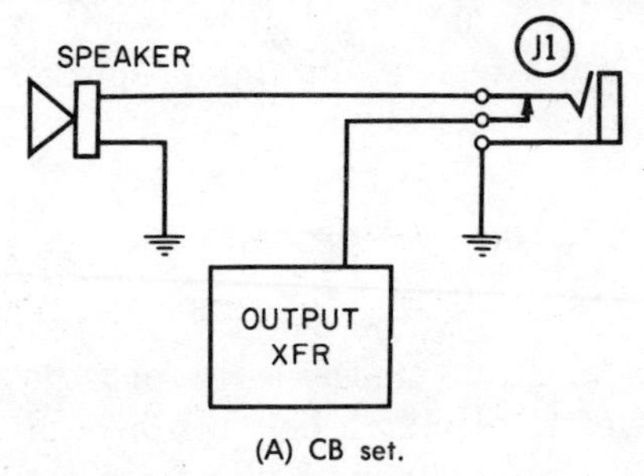

(A) CB set.

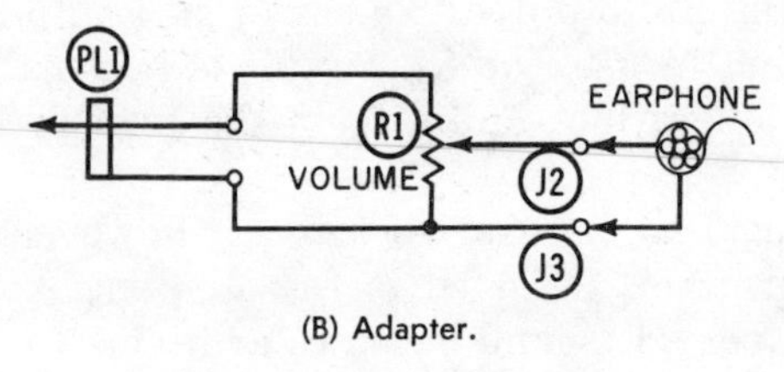

(B) Adapter.

Fig. 6-2. Schematic diagrams.

Actually, the volume control on the CB set can be adjusted for earphone level. The potentiometer on the adapter, however, is added for several reasons; without a separate earphone control, volume in the earphone may be critical to adjust. Since the earphone requires little audio power, the regular volume control might cause blasting, even when opened just slightly. R1 also serves as a load to protect the audio transformer when earphone volume is set at low listening levels. Finally, the potentiometer permits a preset audio level; you can plug the adapter into the CB set without a painful surge of sound in the earphones.

Any popular earphone may be used with the adapter. A typical example is a permanent-magnet (or dynamic) type with an impedance ranging anywhere from 1000 to 5000 ohms. Low-impedance phones may similarly be used. Although a single earphone is shown in the illustrations, the circuit will easily provide enough power for one or several pairs of earphones.

CONSTRUCTION

Assembly of the project is done in two phases—constructing the small metal case and adding jack J1 to the CB set.

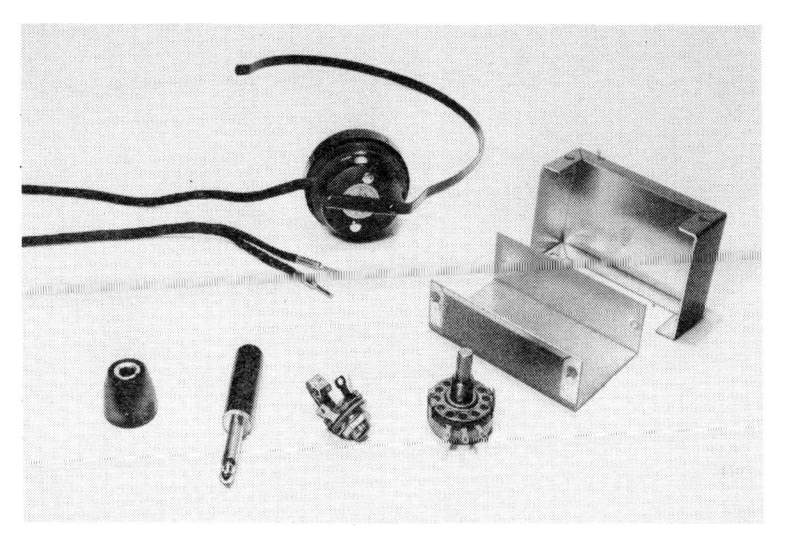

Fig. 6-3. Parts for the adapter.

The basic parts are shown in Fig. 6-3. Note that two phone tip jacks are used on the metal case to receive the earphone connections (Fig. 6-4). If your phones have a different style connector, simply use a matching jack instead of the tips. In any case, neither lead from the earphone should make contact with the metal case. The same holds true for the incoming leads from the CB set; neither makes ground contact with the case. Note, too, how these leads connect to the lugs of control R1. If the control works backwards (that is, rotating the knob of R1 counterclockwise increases earphone volume), reverse the two leads where they connect to the control lugs.

Next, add jack J1 to the CB set (Fig. 6-5). Find an empty spot for mounting, either on the front panel or at the rear. You will have to find one lead in the transceiver to make the proper jack connections; this is the hot voice coil lead to the speaker. Inspect the two wires soldered to the speaker terminals. If it is obvious that one goes directly to chassis ground, the remaining one is the correct one; otherwise, an ohmmeter is required. With the transceiver off, but in the receive position, the proper lead should read about 1 ohm to chassis ground. (The wrong lead shows zero ohms to ground.) Once the correct one is identified, remove it from the speaker terminal and resolder jack J1. It may be necessary to extend this lead with a piece of hookup wire to make it reach the jack. Be sure to insulate the splice with some tape.

Add a length of wire from the free speaker terminal and run it to the appropriate lug on the jack. After this is completed, one lug on the jack should remain unwired; this is the ground connection on the jack which needs no separate lead to chassis ground. When the jack is mounted, the circuit ground occurs when the body of the jack is tightened against the transceiver chassis.

OPERATION

Don't plug in the adapter for the first trial. Turn on the transceiver—sound should be issuing from the speaker with the regular volume control in normal position. This assures that wiring to J1 is correct. Now turn potentiometer R1 on

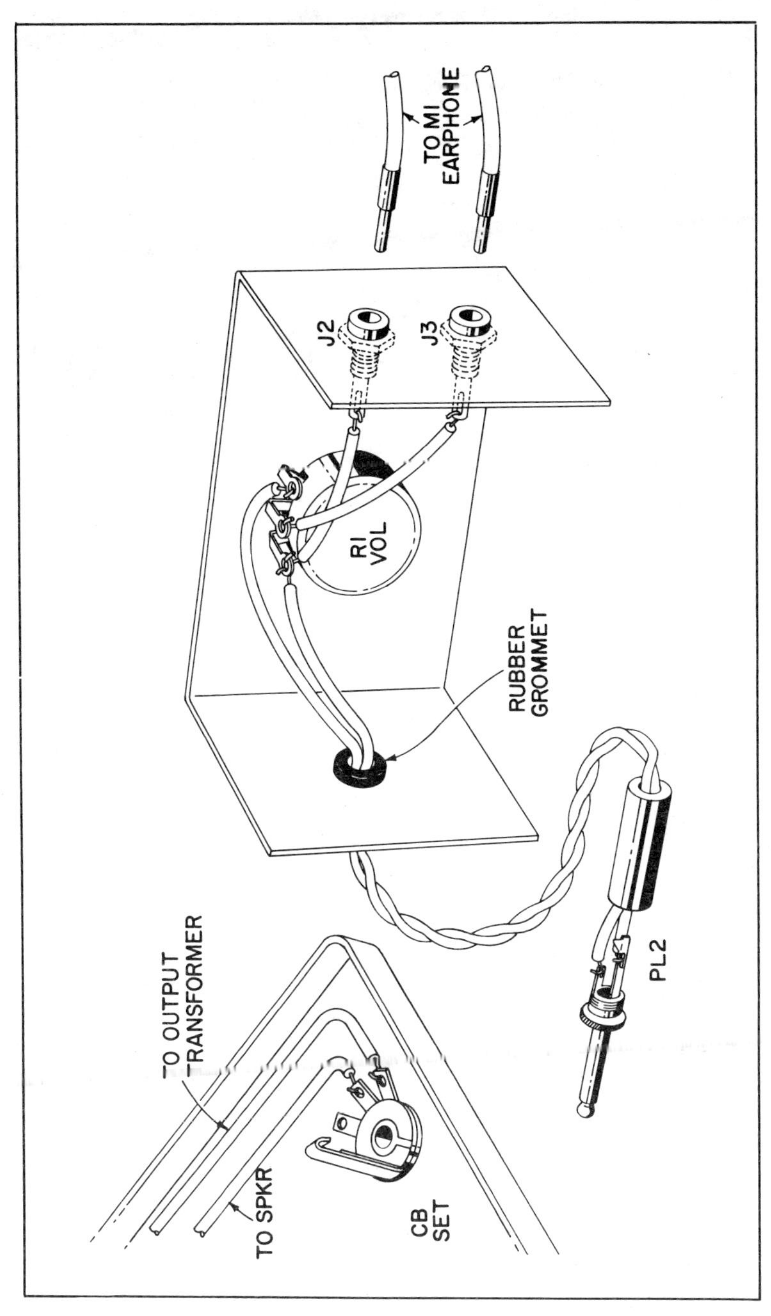

Fig. 6-4. Pictorial diagram.

Fig. 6-5. Phone jack mounted in the CB set.

the adapter to full counterclockwise and plug PL1 into jack J1. As this is done, the speaker volume should cease. While monitoring the earphone, slowly increase the volume by adjusting R1 until a comfortable level of audio is achieved.

The unit is now ready for regular service. However, be sure to leave the regular volume control at its usual setting.

PARTS LIST

Item	Description
R1	50-ohm potentiometer.
J1	Phone jack, closed-circuit type.
J2, J3	Phone tip jacks.
PL1	Phone plug.
M1	Earphone(s).
Misc	Aluminum case, 3″ × 2″ × 1″; rubber grommet; knob.

7

Remote Speaker

One of the simplest ways to increase the versatility of a CB unit is to add a remote speaker (Fig. 7-1). It can carry the sound of an incoming message up to about 100 feet from the main operating position. If you're working in a basement workshop, for example, you won't miss calls being received on the rig upstairs. One application has been in a mobile unit at a construction site. Although the main speaker was mounted in the car, the remote unit could be extended to the site for convenient monitoring. This is especially useful with transistorized transceivers which may be left on safely for hours without seriously discharging the car battery.

CIRCUIT DESCRIPTION

Audio for the remote speaker is picked up from the output transformer of the set exactly as in the preceding chapter. Audio travels over a long interconnecting cable and is fed to the speaker. A wirewound potentiometer in series with one leg of the cable permits remote volume adjustments. The schematic is shown in Fig. 7-2.

CONSTRUCTION

The first part of assembly, adding a jack to the CB rig, follows the identical procedure already described for the headset adapter. Use the same instructions and illustrations for mounting and wiring jack J1.

The potentiometer is mounted at any convenient point inside the speaker cabinet (Fig. 7-3). The thicker the wire of the interconnecting cable, the less power loss there is between the main and remote speakers. No. 20 wire is adequate for runs up to about 50 feet; use No. 18 (lamp cord), or larger, if the distance is much greater.

Fig. 7-1. Remote speaker hooked up.

The circuit will operate like the headset adapter; when the plug is inserted into the jack on the set, the main speaker automatically turns off. While this is desirable for headphone operation, it can be a limitation for remote-

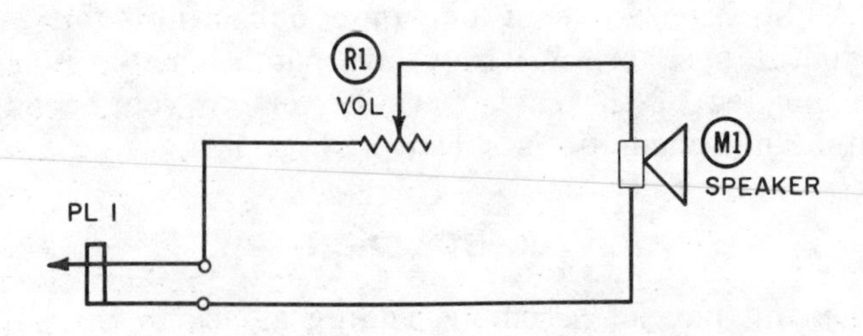

Fig. 7-2. Schematic diagram of remote speaker.

speaker work. It is usually more convenient to have both speakers (main and remote) in simultaneous operation. Converting to this function is possible with the addition of a single jumper wire. Solder a short piece of hookup wire to two terminals on phone jack J1, selecting the lugs that already have wires connected to them (the other remains unsoldered, as before).

Fig. 7-3. Rear view of the remote speaker.

OPERATION

The hookup effectively places main and remote speakers in parallel, and each receives audio power whenever the set is on. Adjusting the overall volume is done by the main volume control on the transceiver and the potentiometer on the remote speaker. It is necessary, however, for the main volume to be left on its normal setting (or slightly higher) for sound to reach the remote speaker. Final settings are easily determined by a few minutes of experimentation. Once they are found, you will be able to reduce sound in the remote speaker without affecting the level at the main operating position.

PARTS LIST

Item	Description
R1	100-ohm potentiometer, wirewound.
PL1	Phone jack.
M1	Speaker, any diameter, 3.2- or 8-ohm voice-coil.
Misc	Two-conductor speaker cable, No. 20 wire or larger; knob, cabinet for speaker.

8

Field-Strength Meter

No other single instrument is as valuable during CB installation as a field-strength meter (Fig. 8-1). Without one, tuning the set into an antenna is largely guesswork. The instrument can be considered a miniature receiver that responds to changes in signal strength, much like an S meter. And after installation is complete, the instrument provides continuous checks on proper operation of the rig at periodic intervals.

The meter described here overcomes a disadvantage found in many simpler-type units. Equipped with a transistor amplifier, high sensitivity permits its use many feet away from the antenna. It avoids erroneous readings that may occur with meters that must be held closer than about 8 feet from the antenna. The problem is that the meter itself becomes part of the antenna system and distorts the readings Also, the antenna radiates two fields of energy—one electromagnetic, the other electrostatic. It is the electrostatic field that carries the radio signal great distances. Close to the antenna, the meter can respond to the electromagnetic field and give false results. These problems are eliminated in this version of the instrument.

Another aspect of the field-strength meter described here is frequency coverage. The most important range, of course, falls in the 27-MHz band. Through the addition of a single switch, however, the same unit can be made to measure output of any transmitter radiating on frequencies from 3 to 30 MHz. This feature greatly extends the utility value of the device for possible use in the future with ham radio equipment. A three-position bandswitch on the front panel selects one of three ranges (in megahertz)—3-8, 8-16, and 16-30. For CB work, the switch is placed in the high band to cover the desired 27-MHz frequencies.

Fig. 8-1. A field-strength meter.

CIRCUIT DESCRIPTION

When a signal enters the antenna of the field-strength meter (Fig. 8-2) it is introduced to a tuned circuit that consists of coil L1 and two-section capacitor C1. The bandswitch (SW1) shorts out sections of the coil according to its position. The switch also connects capacitor section C1B into the circuit when it is on the "Band 1" or 3-8 MHz position. (Additional capacity is required for tuning low frequencies.) Thus, the desired band is first selected and the tuning capacitor rotated until the received signal is peaked up, or resonated, in the tuned circuit. Next, the signal encounters diode D1, where it is changed from alternating to direct current—the type needed by the meter. Capacitor C2 smooths the signal to pure dc.

The transistor (X1) operates as a dc amplifier, taking the weak signal applied to its base and building it up to suitable strength. Meter M2 responds with a reading that varies in step with the strength of the received signal. In this type of circuit the meter does not necessarily double its reading when rf power from the transmitter doubles; it tends to read somewhat higher for small increases in rf energy. This is not a disadvantage, however, since relative changes in signal strength are indicated. During transmitter tuning, the meter need only be observed for the highest reading and not for any precise value of meter current.

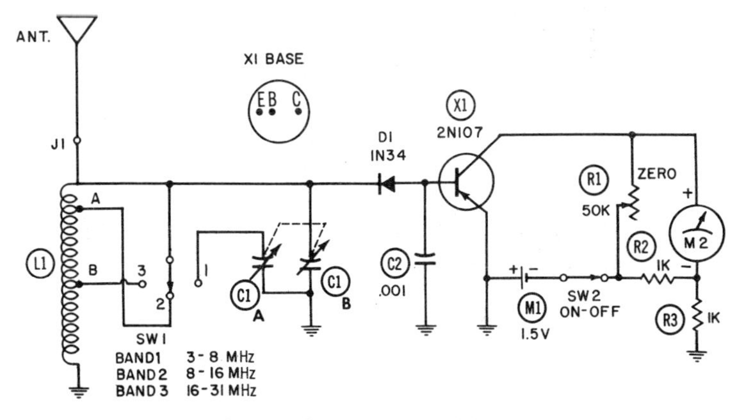

Fig. 8-2. Schematic diagram of the field-strength meter.

Another circuit feature, control R1, is introduced to overcome a possible disadvantage of the transistor. During warm days especially, a tiny amount of current may flow through the transistor and cause a small meter reading even when no transmitter signal is present. The control permits the operator to cancel out this current and adjust the meter to zero each time the meter is used. An On-Off switch connected to this control disconnects power to the circuit when it is not in service.

CONSTRUCTION

As visible in Fig. 8-3, one-half of the aluminum case is cut and drilled for the various front-panel components.

There is a reason for using a large meter; it makes readings visible from a distance.

Note the tuning capacitor; this is a standard component which contains two moving-plate sections that are readily identified by their size. The smaller section will be wired as C1A; the larger as C1B. Such capacitors are commonly sold as replacement units for the variable tuning capacitor used in the standard tube-type a-m table radios. It should be the miniature kind with a frame that measures approximately 1⅜″ × 1⅝″ × 1⅞″. The mounting arrangement is usually through two tapped holes in the bottom of the frame. The illustrations show that the bottom surface of the capacitor rests against the right-side panel of the cabinet with the tuning shaft protruding through the front panel. The detail in Fig. 8-4 reveals an important precaution to observe while fastening the capacitor in place. Pointed out is one of two mounting-screw heads. These must be short screws (usually 6-32) that will not thread too far into the capacitor and short-circuit the fixed plates. Use ¼-inch screws and place several washers under their heads to limit their travel into the capacitor frame.

Most of the transistor amplifier is wired on a small piece of perforated board, as shown in Fig. 8-5. Notice that it

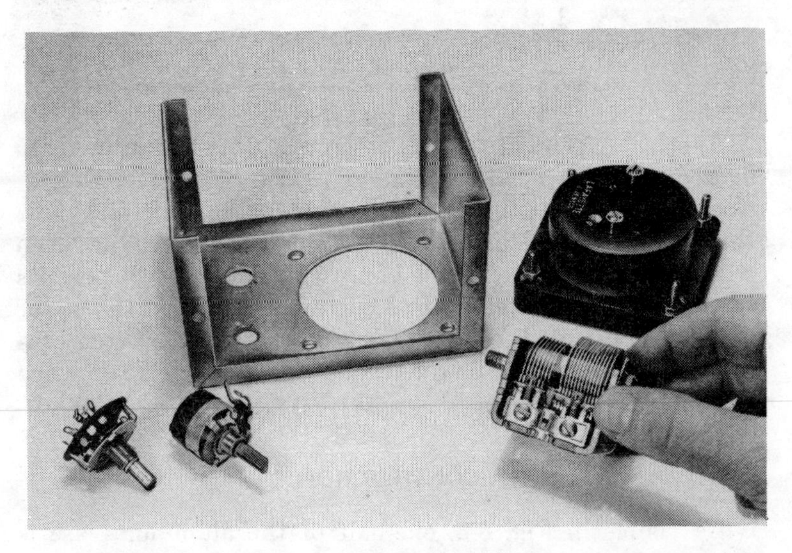

Fig. 8-3. Case and major components.

is mounted by means of the meter-terminal screws. Once these screws are removed and corresponding holes drilled in the board, the board is placed over the meter terminals and the screws replaced. Before this last step, put two solder lugs under the screw heads. Later, when the five-lug terminal strip is installed, these solder lugs are soldered directly to the two outer lugs of the terminal strip.

Fig. 8-4. Location of capacitor-mounting screws.

The battery is held by a strip of scrap metal fastened according to Fig. 8-6. If you don't wish to solder wires directly to the battery as was done in this model, use a battery holder made for such purpose. The drain on the battery is very small, and it can be expected to last for six months or more.

The completely wired unit is shown in Figs. 8-7 and 8-8. The binding post for the antenna (J1) must be one of the insulated types. The post comes with two fiber washers which insert into a hole drilled in the cabinet. One of the washers has a ridge, or shoulder, to center the metal shaft of the post within the hole. While mounting the post, be sure the shoulder washer fits into the rim of the hole and prevents the shaft from shorting against the metal cabinet.

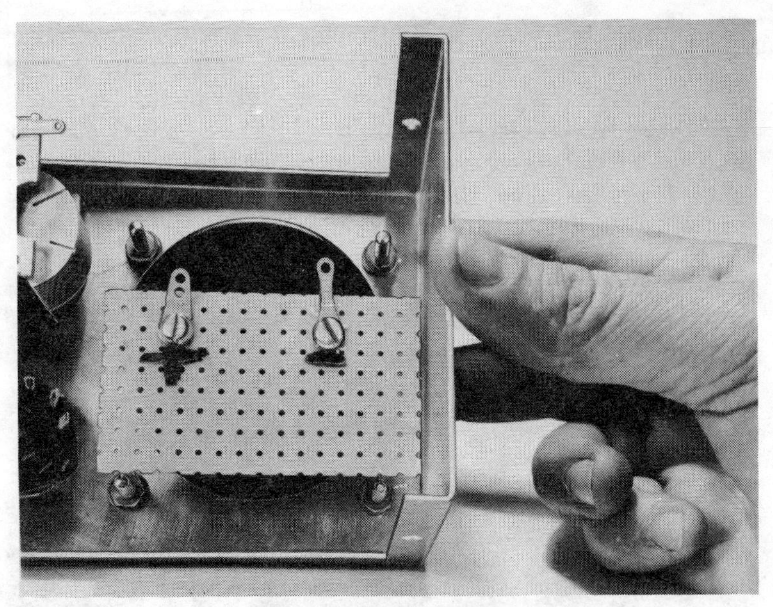

Fig. 8-5. Mounting the board.

Another key point is circuit ground. In the original model, a single wire from the center lug of the terminal strip runs to a ground lug already a part of the tuning

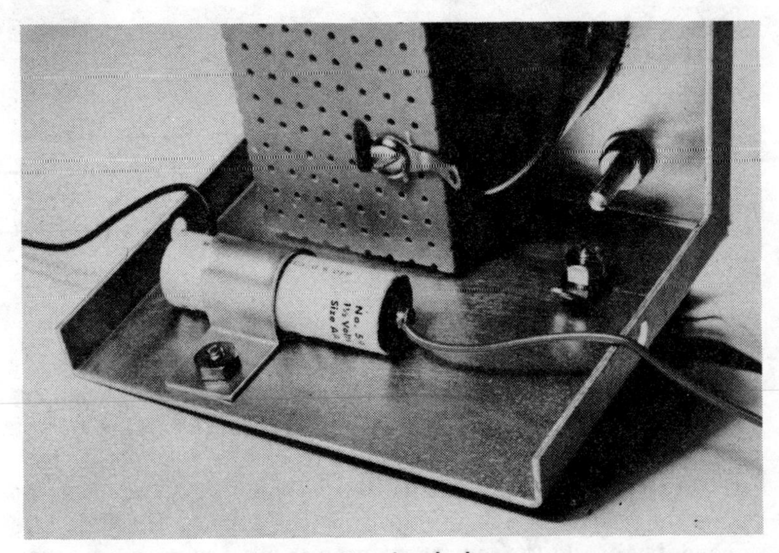

Fig. 8-6. Mounting the battery.

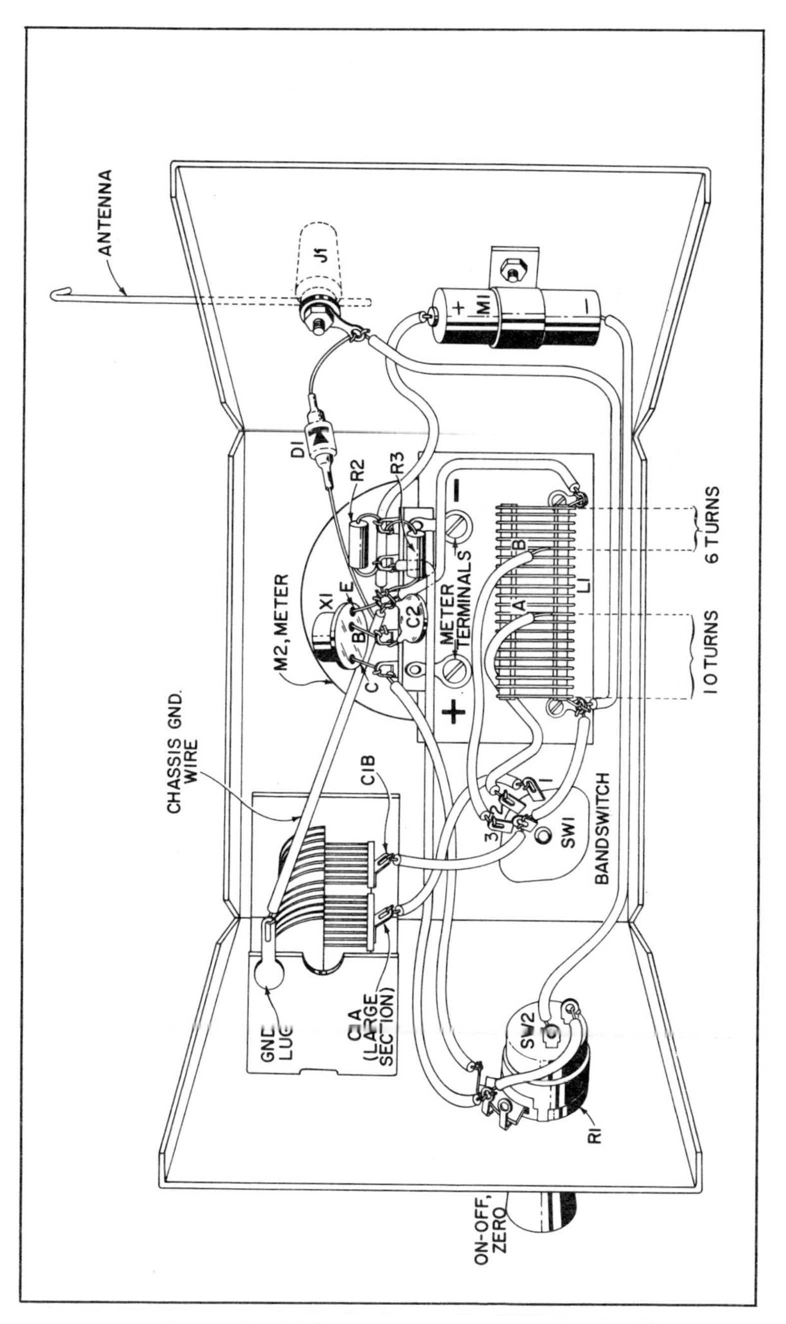

Fig. 8-7. Pictorial diagram of the field-strength meter.

capacitor frame. If your model has no such lug, fasten the ground wire to some point that makes contact with the metal case. A solder lug placed under the nut that holds the battery strap in place is one alternative.

Fig. 8-8. The completed project.

After wiring of the circuit is complete, the final step is to find the correct tap-in points for wires marked *A* and *B* on coil L1. Although the pictorial gives the number of turns in from the edge of the coil, the figures might not be exactly the same for your unit. From one circuit to the next, small layout differences in wiring affect the placement of the wiring. The technique for finding the precise points is to slide one coil-tap wire across the turns in an effort to locate the proper point. This must be done while a fairly strong signal is received. Since you may not have access to signals on the lower bands at the present time, we'll concentrate on the adjustment of the high band (16-31 MHz) on which the CB frequencies are found.

Begin by soldering wire *A* to the tenth turn on the coil as shown in Fig. 8-2. Now turn the tuning capacitor knob on the front panel until the capacitor plates are completely open (unmeshed). Next, turn the knob one-quarter turn in the opposite direction and leave it there. (You can judge the quarter-turn by viewing the pointer or set screw on the knob as it is rotated.)

What is being done during this phase is calibrating the capacitor. In the first step, the fully open plates correspond to 31 MHz. Then, the plates are closed one-quarter turn to bring them into the 27-MHz position, and, finally, the coil is brought into range. Turn the power switch on and also put your CB set on transmit. (You may wish to use a dummy load on the rig during this step to avoid transmitting an interfering carrier while calibrating the coil.) Grasp wire *B* by its insulation and move the bare tip along the coil, starting from the right end. At one point during this process the meter pin should rise sharply to indicate that the instrument is responding to the radio wave being emitted by the CB set. Solder the wire directly to this coil turn. It should be somewhere in the vicinity of six turns from the coil end, as shown in the pictorial.

Wire *A* can be calibrated at a future time when low-frequency signals are available. The process is basically the same as just described. The position of wire *A* determines where the midband will fall.

OPERATION

The antenna used on the field-strength meter influences the sensitivity of the unit. In most instances a piece of stiff wire approximately 10 inches long will suffice (see Fig. 8-1). This may be extended for added pickup, especially for the low bands.

A valuable application for the field-strength meter is in tuning up a mobile rig. With the transmitter on, hold the meter within a foot or two of the antenna. The bandswitch should be set to the high band (knob clicked fully clockwise) and the tuning capacitor slowly varied until the meter indicates signal pickup. This should be done slowly to avoid pinning the needle in case a strong rf field is present. Now move away from the mobile unit until you can peak the needle to a reading near the low end of the scale. Placed about 10 or more feet from the mobile antenna, the meter should indicate the effects of small changes in transmitter tuning adjustments. If the meter is too sensitive and continuously reads full scale, simply shorten its antenna or detune the variable capacitor to reduce the reading. After all adjustments are complete, you can use this same setup as a reference for future checks. With the meter and mobile antenna in the same relative positions, a lower reading in the future suggests trouble in your mobile transmitter or antenna system.

PARTS LIST

Item	Description
R1	50K carbon potentiometer with switch.
R2, R3	1K resistors, ½ watt.
C1A, C1B	Dual-section variable capacitor, broadcast type (C1A is large section, C1B is small section).
C2	.001-μF disc ceramic capacitor.
D1	1N34 germanium diode.
J1	Binding post, insulated type.
SW1	Single pole, 3-position rotary switch.
SW2	On-off switch (part of R1).
M1	1.5-volt size AA battery.
M2	0- to 1-mA dc milliammeter, 2½-inch face.
X1	2N107, GE-2, SK3003, or HEP250 transistor.
L1	Coil, ¾" diameter, 16 turns-per-inch (B&W Miniductor No. 3011). See text.
Misc	Aluminum case 5" × 3" × 4"; perforated board 1¾" × 3¼"; three knobs; 8" length of stiff wire for antenna; 5-lug terminal strip; two ground lugs; scrap metal ½" × 2" for battery strap.

9

TVI Trap

CB transceivers are equipped with internal filters designed to reduce interference to tv reception. The filter tends to lessen radiation of the second harmonic (54 MHz), which usually disturbs tv Channel 2. There is, however, a type of interference termed fundamental blocking. It stems from the inability of some tv sets to reject the CB carrier on 27 MHz. The problem is aggravated when CB and tv antennas are in close proximity.

The tvi trap (Fig. 9-1) is intended solely to reduce fundamental blocking. Recognizing this type of interference can be done while transmitting on the CB unit and observing the tv set in operation. Disturbance to sound or picture generally occurs over several channels, rather than being restricted to Channel 2, as in the case of harmonic interference.

CIRCUIT DESCRIPTION

The tvi trap consists of a tuned circuit on 27 MHz placed directly across the two wires of the twin lead from the tv antenna. Television channel frequencies, which range from approximately 54 to 215 MHz, are little affected by the trap. But the trap presents a virtual short circuit to a 27-MHz signal. The interfering signal is thus detoured before it has a chance to enter the tv set.

Fig. 9-1. The tvi trap.

CONSTRUCTION

The trap is assembled entirely within a small aluminum case. In Figs. 9-2 and 9-3, after the terminal strip (TS1) is fastened to one end of the case, determine the spacing of the two lug-type terminal strips. (Use capacitor C1 as a spacing guide.) The two lugs of the capacitor are soldered to insulated lugs on the terminal strips. As shown in Fig. 9-3, the terminal-strip lugs are bent into a horizontal position to form a flat support for the capacitor.

Winding the coil is next. Clean the end of a length of No. 20 enameled wire and proceed to wind 22 turns around a ¼-inch coil form. An ordinary pencil is close enough in diameter to serve as the form. When the coil is complete, slide it off the form and solder the two clean ends. A piece of twin lead, about 6 inches in length, is fastened under the

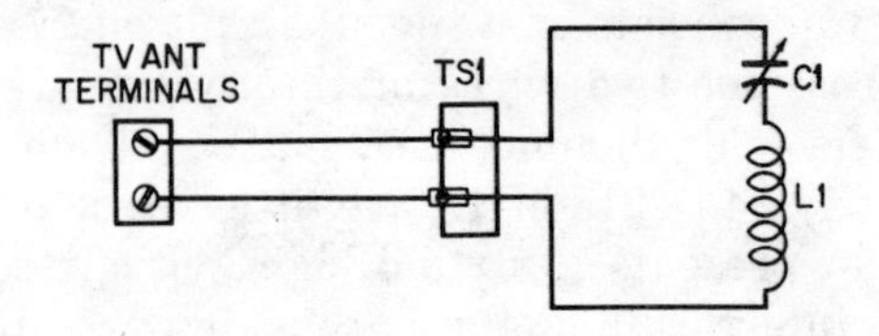

Fig. 9-2. Schematic of the tvi trap.

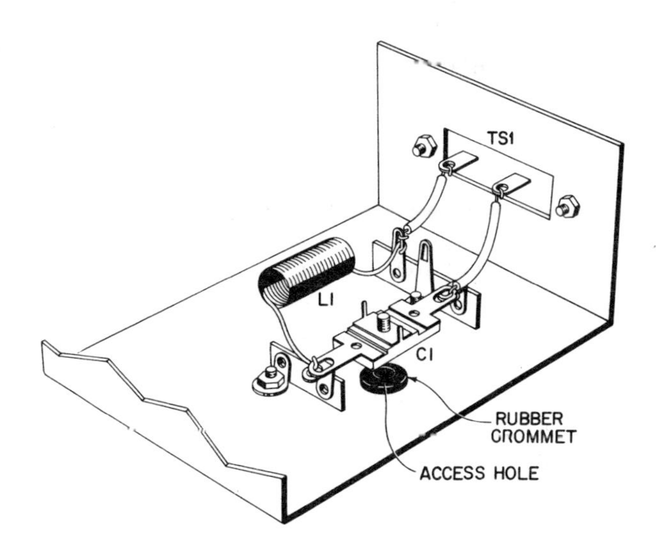

Fig. 9-3. Pictorial diagram of tvi trap.

Fig. 9-4. Tvi trap with cover removed.

screws to TS1. The completed unit, with cover removed, appears in Fig. 9-4.

OPERATION

The free end of the 6-inch twin lead from the trap is placed under the antenna terminals of the tv set (the original twin lead from the tv antenna is left in place). While viewing or listening to the interfering signal from the CB set, begin tuning capacitor C1 through the access hole in the aluminum case, as in Fig. 9-5. (The job is least critical if a nonmetallic screwdriver is used.) Adjust the tuning screw on the capacitor for minimum interference.

Fig. 9-5. Tuning C1.

Due to differences in wiring, your unit may not tune precisely to 27 MHz. One suggestion is to slightly pull apart the turns of the coil. After each trial, adjust the tuning capacitor over its full range. If this produces no effect on the interference, try increasing the length of the coil by two or three turns for each trial.

The performance of the trap might be improved by providing it with a good electrical ground. If the tv set is a transformer type, run a short wire from one of the screws on the aluminum case to a screw on the tv chassis. (Transformer-type tv sets may be identified by the fact that they are labeled for ac operation only.) The ac-dc sets may cause a shock hazard if the chassis is used for ground purposes. In this case, ground the tvi trap to a nearby cold-water pipe or under the screw that holds the cover plate of the ac wall outlet.

Again the trap is effective for fundamental blocking only. If interference is from harmonics, be sure to check whether the filter in the CB set is properly adjusted; the procedure is usually described in the unit instruction manual.

PARTS LIST

Item	Description
C1	3 to 30-pF trimmer capacitor (compression type).
L1	Coil, 22 turns No. 20 enamel-covered wire, 1/4" dia.
TS1	2-screw terminal strip.
Misc	Rubber grommet; aluminum case, 3" × 1" × 2"; two 2-lug terminal strips; 6" tv twin lead, 300 ohms.

10

Line Filters

A major path for interference into the CB receiver is via the ac line. Numerous household appliances with motors or sparking contacts may generate noise and impress it onto the line. Typical of these units are mixers, refrigerators, and vacuum cleaners. Fluorescent lamps operate in rapid charge-discharge fashion that tends to produce a train of noisy rf pulses on 27 MHz. This results in hash, which often obliterates weak-signal reception in a CB receiver. Much of it can be cured with suitable line filters. Three types are shown in Fig. 10-1—from simple bypassing to the more elaborate choke-type filter. To varying degrees, each can short-circuit noise energy before it enters the receiver.

Since interference can be radiated from the ac line and picked up directly by the CB antenna, the best place to begin is at the source. A noisy appliance not only generates hash in its wiring, but the emerging line cord and associated house wires may act like an antenna for these frequencies. Filtering inside the appliance helps to kill the objectionable energy before it travels out of the appliance enclosure into the line.

CIRCUIT DESCRIPTION

Fig. 10-1 shows the three principal techniques. In the single-capacitor bypass circuit a .01-μF ceramic-disc capaci-

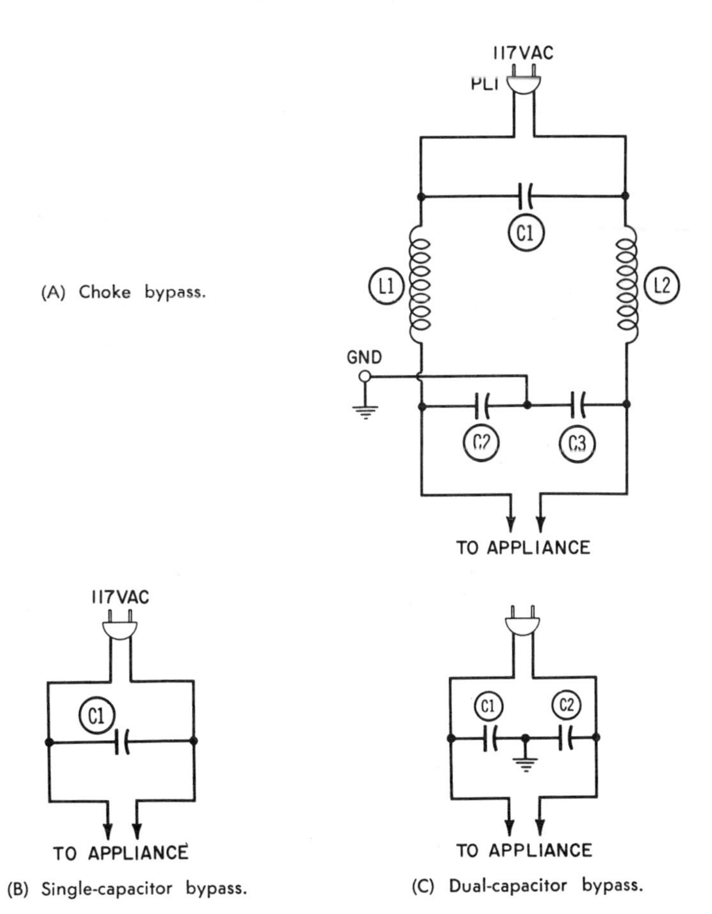

(A) Choke bypass.

(B) Single-capacitor bypass.

(C) Dual-capacitor bypass.

Fig. 10-1. Line filters.

tor is connected across the ac line. It forms a low-impedance path for noise frequencies, while leaving the 60-cycle ac power undisturbed. The second circuit is a dual capacitor bypass circuit which employs two .01-μF capacitors on the line and a separate ground connection. The system gives somewhat better results than the single bypass.

The most elaborate unit is the choke-bypass combination. Aside from three bypass capacitors, two rf chokes are included to offer great opposition to the flow of noise frequencies from the appliance into the ac line. The chokes are constructed to present little obstruction to the normal flow of ac power.

CONSTRUCTION

Only details on the choke-bypass filter are given in this section since there is no actual assembly for the simpler bypass capacitors. (Practical installation details on the capacitors are given later.)

The basic components for the device are illustrated in Fig. 10-2. The terminal strips should be of the type with grounding feet at the outer ends. These provide electrical grounds for several points in the circuit. The external ground terminal is a ½-inch machine screw on the side of the metal cabinet. Two units threaded on the screw enable the ground wire to be attached, as in Fig. 10-3.

The assembly of the two choke coils is not critical. As long as they have the approximate number of turns and diameter given in the parts list, they will provide the proper action. Just be sure that the wire is No. 18 enamel covered so that no heating occurs as the appliance draws current through the unit. The coil forms in the model shown are

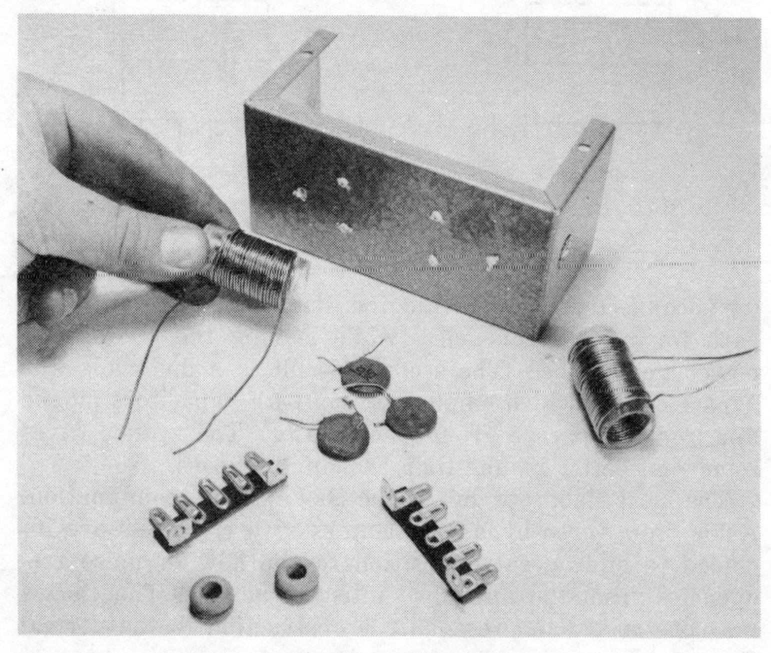

Fig. 10-2. Parts layout.

plastic types with a tapped hole in the bottom surface. This lends to simplified mounting with 6-32 machine screws installed through the bottom of the cabinet (Fig. 10-4). A solid wood dowel may also serve for the forms.

OPERATION

One approach for treatment of a line-noise problem is a trial with the simple bypass capacitor. It is most useful

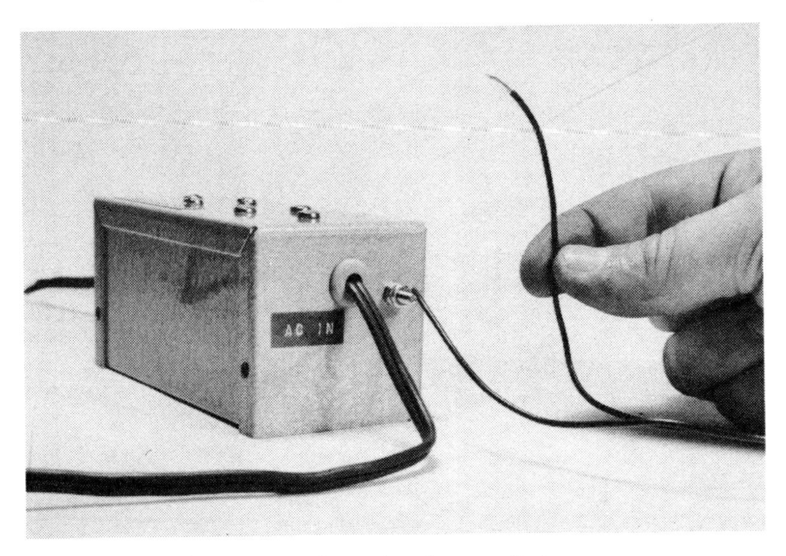

Fig. 10-3. Connecting the ground wire.

when the appliance has a two-wire ac cord and no easily accessible ground connection. The basic layout is given in Fig. 10-5. The capacitor is wired directly across the line, and the best location is inside the appliance enclosure. After tracing the point to which the ac wires connect, the capacitor is soldered in place. Extreme care must be exercised to prevent short circuits. Ample use of black plastic tape on the capacitor leads should prevent the problem. If space for capacitor mounting is limited, one of the smaller 600-volt ceramics may be used. (Usual voltage rating for a disc ceramic is 1000 volts.)

When it is impossible to install the capacitor inside the appliance, the choke-bypass unit is recommended. Although

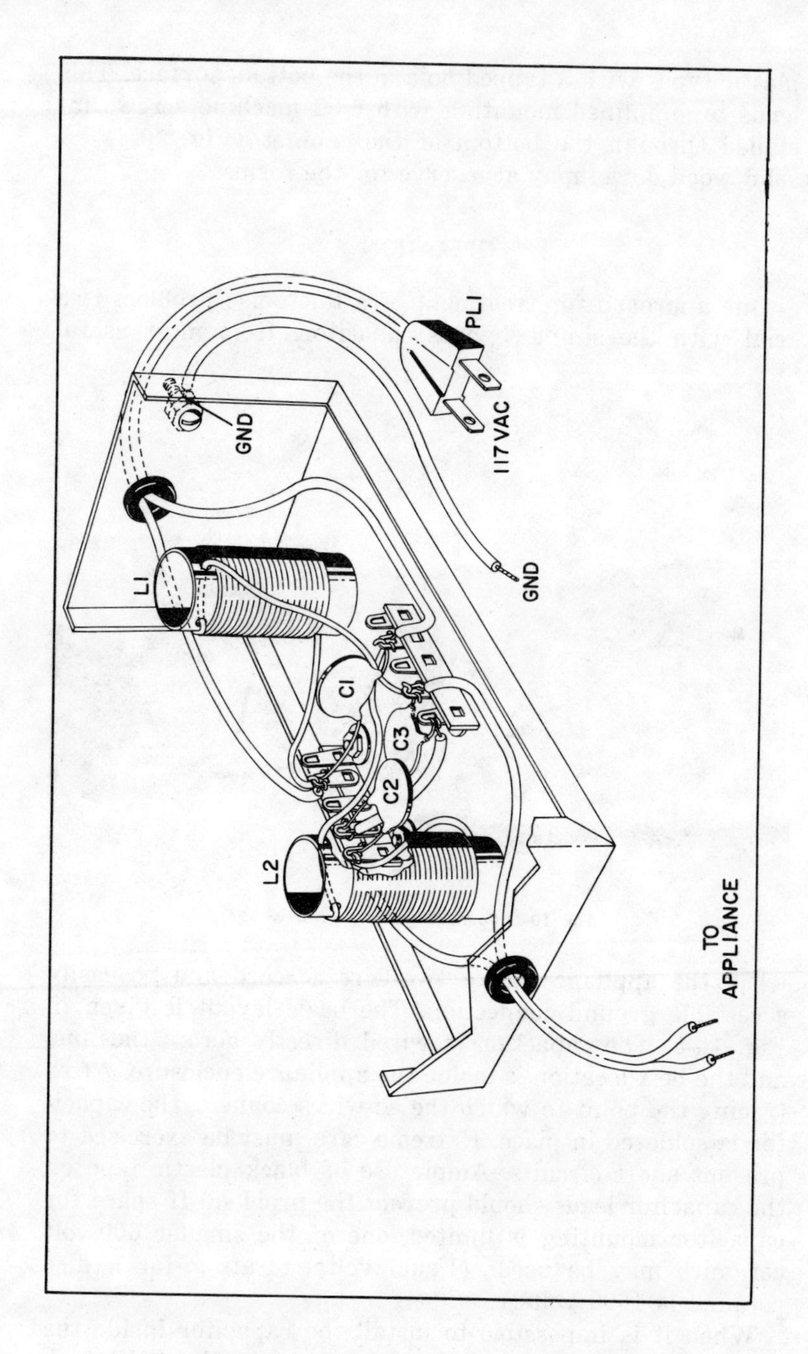

Fig. 10-4. Pictorial diagram of the choke-bypass line filter.

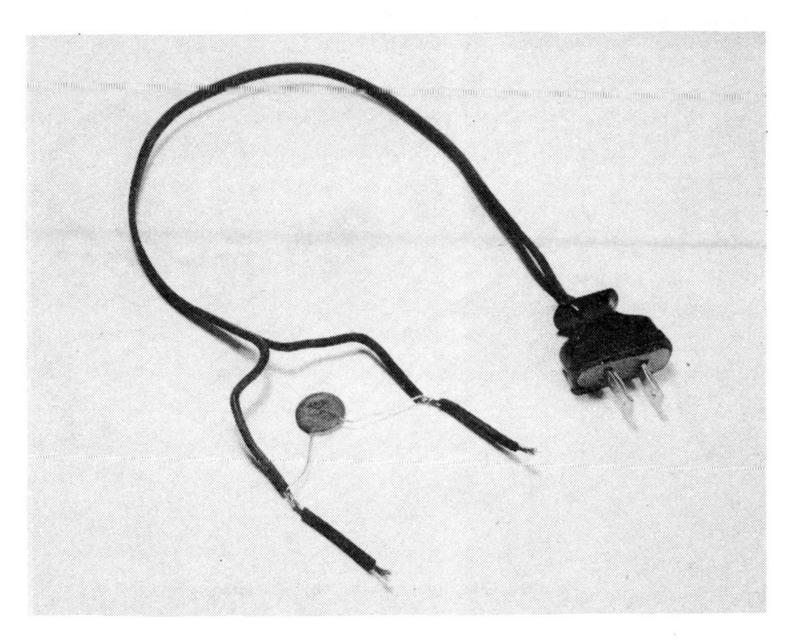

Fig. 10-5. Single-capacitor bypass filter.

it is a more elaborate device, it provides a convenient physical arrangement for hooking into the line.

An important step in using any of the filters described is to insert the line plugs into the ac outlet both ways to discover the position that produces least noise in the CB set. The same trial should also be performed on the CB line cord.

The second filter, the dual bypass, requires a known electrical ground. This is not easily available in small appliances having a two-wire ac cord. It is suitable, however, for three-wire devices. If the appliance has a three-prong plug (often used with an adapter on a two-wire ac outlet), the two capacitors may be installed with a ground to the appliance case. The basic layout is given in Fig. 10-6. Note that the ground leads on the capacitors go to a lug for attachment to a screw on the appliance case.

An application where the dual bypass proves especially effective is reducing the buzz-type noise generated by fluorescent lamps mounted in a ceiling fixture. Before opening the fixture, be sure that all ac power is off by removing the appropriate house fuse.

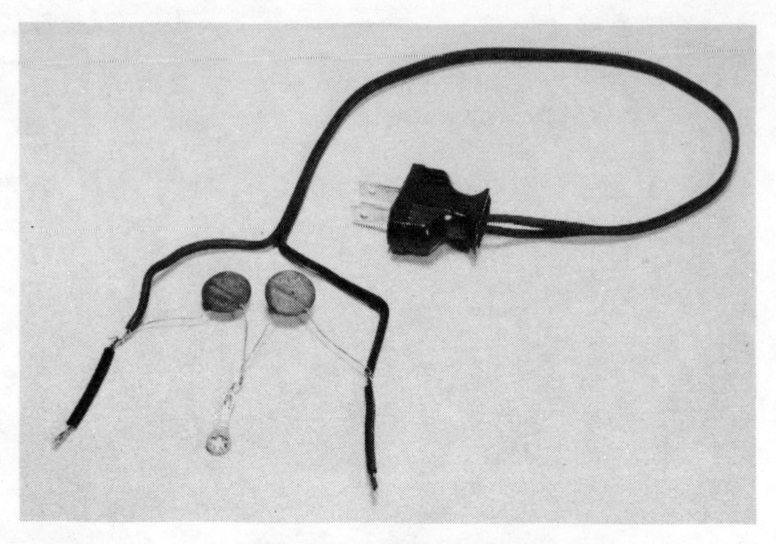

Fig. 10-6. The dual-capacitor bypass filter.

The final filter (choke-bypass) is placed in series with the appliance line cord. It is intended to treat most types of noise but has special value for eliminating hi-fi or tv set interference originating from the CB transmitter. It is possible that radio-frequency energy from the CB set may be impressed onto the house wiring and voice modulation heard on home-entertainment equipment. The filter is best installed in the line, close to the body of the offending appliance or equipment. The ground lead connects to a nearby cold-water pipe or under the screw that holds the cover plate on the ac wall outlet.

When filtering at the source proves ineffective or physically impossible, additional suppression is possible by installing the choke-bypass filter in the line cord of the CB set.

PARTS LIST

Item	Description
C1, C2, C3	.01-μF disc-ceramic capacitor.
L1, L2	Choke coil, 25 turns No. 18 enamel wire wound on ¾″ coil form.
PL1	Ac plug and line cord.
Misc	Aluminum case, 2¼″ × 2¼″ × 5″ approx; two 5-lug terminal strips, outer lugs grounded; 6-32 machine screw, ½″ long with two nuts for ground terminal.

11

Standing-Wave Meter

In order for a transmission line to deliver maximum power into an antenna, correct matching must exist. This is the ideal situation in standard CB systems; a 52-ohm coaxial transmission line feeds an antenna that displays 52 ohms at its input terminals. Any mismatching between line and antenna results in a power loss. This occurs when current through the line fails to be absorbed by the antenna and reflects back through the line to the transmitter. The result is a reduction in available rf energy, since reflected power cancels the original, or incident, power generated by the transmitter.

The relationship between reflected and incident power is swr (standing-wave ratio). If the system is perfect, a theoretical 1:1 ratio exists between maximum and minimum current at any point in the transmission line. A severely mismatched system may have an swr of 10:1, for example. While this is often acceptable in certain multiband antenna systems that utilize low-loss line, it is intolerable in the coaxial-type line employed for CB work.

The practical value of knowing the swr is that you can take steps to reduce it. The closer the system is adjusted to a 1:1 ratio, the more efficient will be the radiation from the antenna. This is the purpose of the device described here and shown in Fig. 11-1. Inserted into the transmission line, it indicates reflected power. As corrective measures are made, the meter is observed for the lowest reading. The instrument will not indicate an absolute value of swr; that is, it will not measure whether the swr is 1:3, etc. Such a circuit is more critical and complex to build. Rather, the device gives relative indications of reflected power which enable adjustments to be made. It may be assumed that as reflected power is reduced, the swr figure goes down with it.

Fig. 11-1. The standing-wave meter.

Practical applications of the meter include antenna adjustments and checking; some CB antennas are equipped with a sliding element that must be fixed into final position by the user. The meter supplies information needed to perform this adjustment—the lowest reading occurs at the correct antenna length. The same process is helpful in the design of homemade antennas. Also damage to a matching element in an antenna can be corrected with the aid of the meter.

CIRCUIT DESCRIPTION

The circuit forms a simple bridge that compares line currents—both incident and reflected. In Fig. 11-2 rf energy from the CB set is introduced to jack J2 and across resistor R1. The resistor is 52 ohms in order to provide a

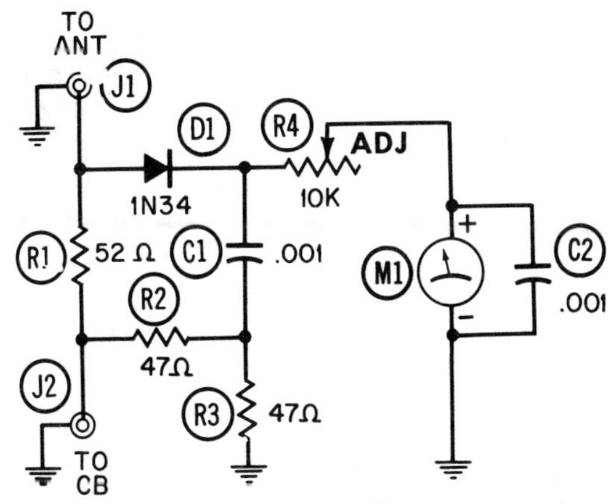

Fig. 11-2. Schematic of the standing-wave meter.

reference voltage. Reflected power enters jack J1 from the antenna and causes a meter indication after signal rectification occurs in diode D1. Potentiometer R4 is provided to bring the meter pin into proper range and prevent off-scale readings.

CONSTRUCTION

Good performance of the meter requires careful construction. As illustrated in Fig. 11-3, two coaxial connec-

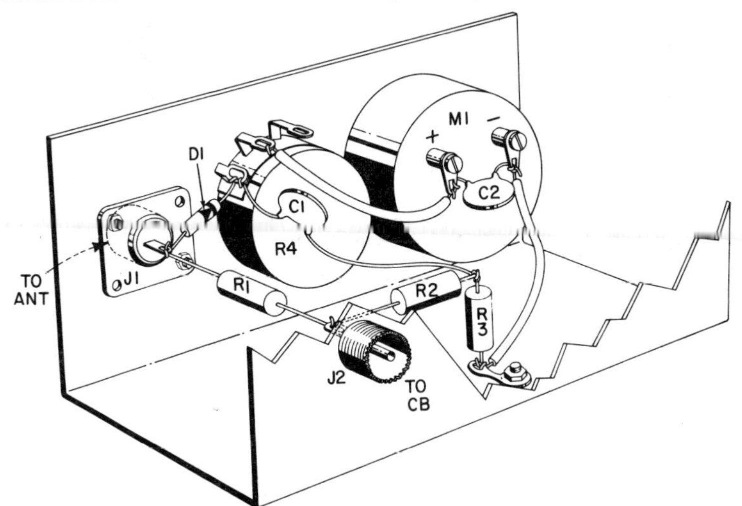

Fig. 11-3. Pictorial of the standing-wave meter.

tors (J1 and J2) are mounted on opposite sides of the aluminum case. When the three fixed resistors are soldered in place, follow the layout shown; keep their leads short and equal in length. Resistors R2 and R3 should be the same distance from the side of the chassis. Note that a solder lug fastened to the bottom of the case provides a connecting point for the lower lead of R3 and a ground for the negative meter lead. The hole for the meter is 1½″ in diameter if the small case called for in the parts list is used.

Since the instrument is used only on an occasional basis, some builders might prefer not to mount the meter, but use an external milliammeter. If so, provide two pin jacks instead of the meter on the front panel of the instrument. Red and black jacks can identify positive and negative meter-lead connections. Any multimeter with a 0- to 500-

Fig. 11-4. Inside view of the standing-wave meter.

microampere or 0- to 1-mA range is suitable for indicating purposes. An overall view of the circuit layout is given in Fig. 11-4.

Coaxial connectors J1 and J2 are not necessarily standard for CB equipment. If your CB unit uses different type plugs and jacks, these should be used on the swr meter.

OPERATION

The instrument is connected by removing the coaxial transmission line from the CB set and plugging it into J1. A short length of line with suitable connectors is run from J2 to the antenna jack of the CB set (Fig. 11-5).

Turn adjustment potentiometer R4 fully clockwise before applying power. This setting is the least sensitive for the meter and protects it against sudden off-scale readings. (Note that the knob works opposite from the usual direction. The meter is most sensitive when R4 is fully counterclockwise, or to the left.) Transmitter power may now be applied. The potentiometer is rotated until the meter pin shows some convenient reading about half-way up the scale.

Fig. 11-5. Meter hooked up to a CB unit.

Using the device is a matter of making antenna adjustments while monitoring the meter for the lowest possible reading. If the pin drops to nearly zero, increase sensitivity with R4 until a comfortable indication is achieved.

When operating a meter of this type, it is not abnormal to see a high meter reading even though the amount of reflected power is very low. The indications are purely relative; the lowest meter current is the objective, as opposed to some particular value of current. Finally, remove the meter from the transmission line after adjustments are completed. The device is not intended for continuous monitoring during the course of regular CB operations.

PARTS LIST

Item	Description
R1	52-ohm resistor, 1 watt.
R2, R3	47-ohm resistors, 1 watt.
R4	10,000-ohm potentiometer.
C1, C2	.001-μF ceramic-disc capacitors.
D1	1N34 germanium diode.
M1	0- to 1-mA dc meter (see text).
J1, J2	Coaxial connectors.
Misc	Aluminum case, 5″ × 2¼″ × 2¼″; solder lug.

12

Output-Power Indicator

The output-power indicator shown in Fig. 12-1 is actually two pieces of equipment housed in a single case. It has a sensitive meter that reads relative power output and a dummy load to provide a proper match to the transmitter. The unit is a handy device for testing, adjusting, and troubleshooting.

Although a No. 47 pilot lamp is widely used for the functions described, a meter is capable of readings that are much easier to view. Even small transmitter adjustments produce a clearly visible swing of the needle. You can tune a transmitter and check for nonoscillating crystals, poor tubes, and other items that may affect rf output.

CIRCUIT DESCRIPTION

In Fig. 12-2 output power from the transmitter is introduced to jack J1. A resistor network comprised of R1 through R4 makes up the dummy load. Since each resistor is 220 ohms, the total resistance in this practical connection is 55 ohms. This value is close enough to 52 ohms to provide the proper match and load for the transmitter. Rf power is dissipated in the form of heat in the dummy load.

The meter circuit is driven by a sampling of rf voltage across the dummy load. The rf energy is rectified by diode D1 and filtered by capacitor C1. The result is smooth dc that is approximately equivalent to the rf voltage from the

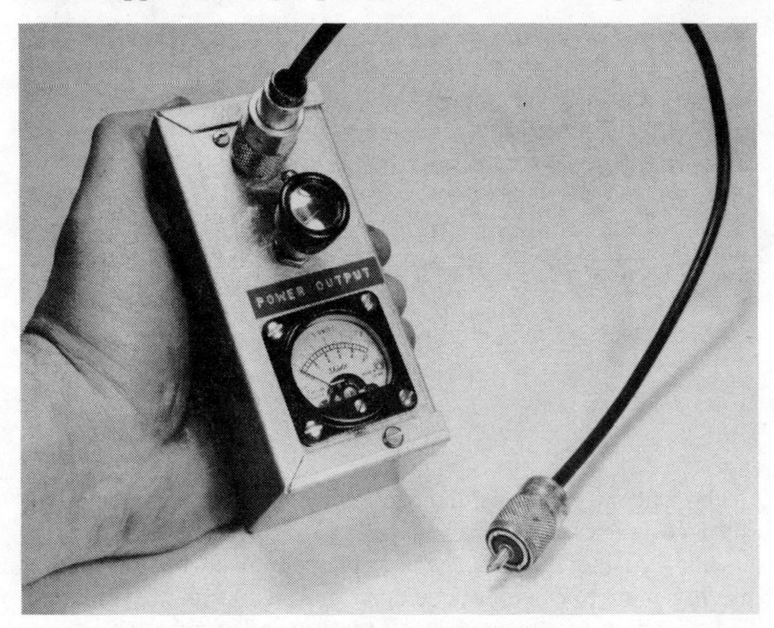

Fig. 12-1. The output-power indicator.

transmitter. The principle of operation is that voltage to the meter varies with output power. The meter reads the dc sampling after it passes through R5, a meter-sensitivity control. The 10,000-ohm resistor (R6) from meter to

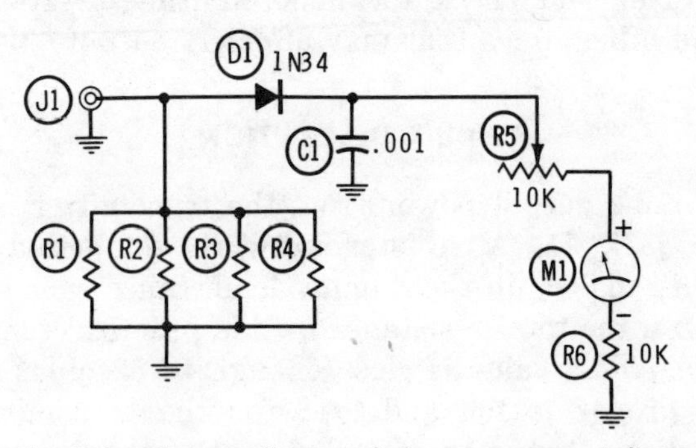

Fig. 12-2. Schematic diagram of the output-power indicator.

ground is a current-limiting component to prevent the needle from rising off-scale.

Since the circuit is essentially an rf voltmeter, it will not read output watts directly. Relative readings, however, are ample enough for checking transmitter operation, and the circuit remains simple and inexpensive to build.

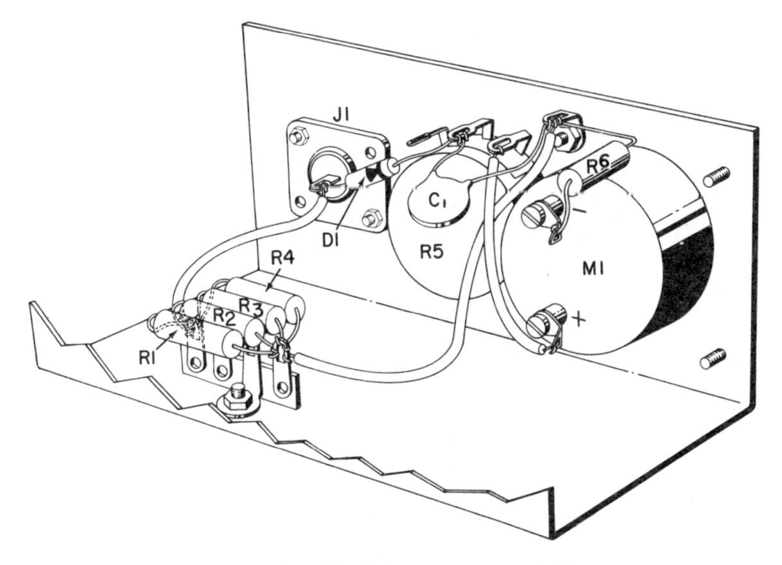

Fig. 12-3. Pictorial of the output-power indicator.

CONSTRUCTION

As shown in Fig. 12-3, most of the major components are mounted to the front panel; the meter, sensitivity control R5 and the input jack. Be sure to install a solder lug under one meter-mounting screw to provide a chassis ground for resistor R6, capacitor C1, and the ground for the dummy load.

The dummy load is constructed separately from the rest of the circuit. The completed assembly of four resistors should lie flat with a slight spacing between each resistor. (The space permits air circulation between the units which become warm in operation.) Otherwise, wiring of the circuit is neither critical nor sensitive to exact placement of leads.

A cable about 1 foot long is now prepared. This is the lead that carries the output energy of the transmitter to the instrument. Although a coaxial connector is shown in this model, use the type that matches your CB output. The cable is cut from 52-ohm transmission line and male plugs installed on both ends. An inside view of the completed unit is shown in Fig. 12-4.

Fig. 12-4. Inside view of the indicator.

OPERATION

The device is set into operation when the coaxial cable is connected between the CB set antenna-output jack and the input connector J1 (Fig. 12-5). To prevent the meter needle from pinning at the top of the scale, turn the sensitivity knob to the full counterclockwise position. Turn on the transmitter and raise meter sensitivity until the pin rises off the zero mark. A setting that places the pin about halfway up the scale is adequate for most purposes.

The transmitter may now be tuned for maximum output as indicated by a high meter reading. One way to increase

Fig. 12-5. Hookup of the indicator to a CB unit.

the value of the instrument is by calibrating the sensitivity control. Assuming that the CB transceiver is performing perfectly, set the sensitivity knob so that the meter is at half scale. If this knob position is marked it serves as a future reference. Lower future readings with the knob on the same mark reveal a problem of low rf output.

PARTS LIST

Item	Description
R1, R2, R3, R4	220-ohm, 1-watt resistors.
R5	10,000-ohm potentiometer.
R6	10,000-ohm resistor, ½ watt.
C1	.001-μF ceramic-disc capacitor.
D1	1N34 germanium diode.
J1	Jack, coaxial connector.
M1	0- to 1-mA dc meter, 1½" dia.
Misc	4-lug terminal strip; solder lug, aluminum case, 5" × 2¼" × 2¼".

13

CB Scope Adapter

Few CB'ers own an oscilloscope, but this test instrument is often available through a fellow CB club member or neighborhood experimenter. If it is properly used, a scope provides an excellent means of analyzing important aspects of an rf signal. It indicates modulation percentage, relative output, hum, distortion and overmodulation.

Most scopes, however, cannot display a 27-MHz carrier signal fed into a vertical amplifier. The frequencies involved are far too high. An alternative is to impress the signal directly on the scope's vertical plates. This works, but the signal display is usually low in height and barely readable. The answer is the use of a scope adapter (Fig. 13-1). The one described here is designed expressly for CB work and presents a wide, clearly visible indication. Also covered in this chapter are patterns you can get and how to interpret them.

CIRCUIT DESCRIPTION

The adapter shown in Fig. 13-2 is a coupling device between CB output and scope input. Rf energy is introduced to jack J3 where most of its power is dissipated by resistor network R1 through R4. Since the scope needs negligible

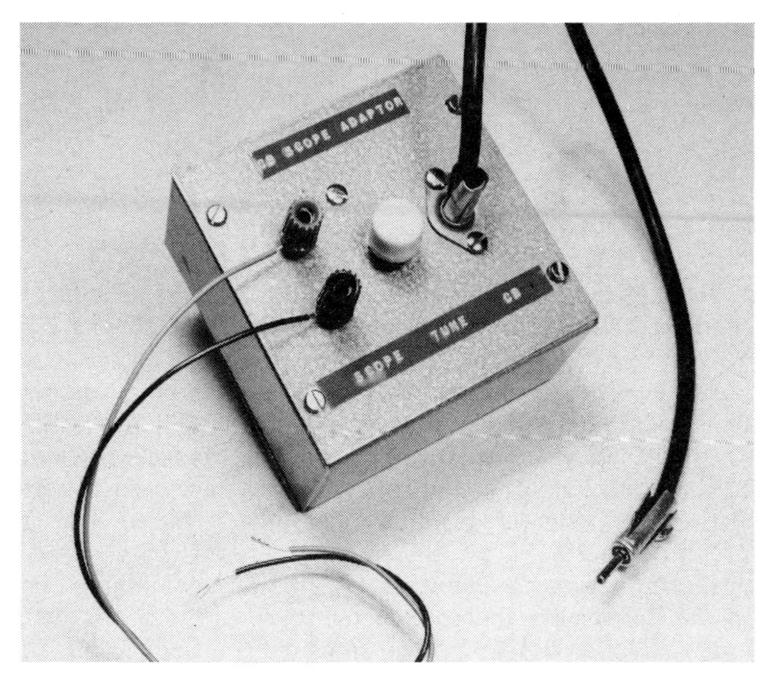

Fig. 13-1. The CB scope adapter.

power from the transmitter, the resistors safely convert the energy into heat.

The scope input requires far higher voltages than those available from the low-impedance output of a CB transmitter. The required step-up occurs in coil L1. Winding B of the coil couples rf energy to winding A, which is part of a tuned circuit. The 27-MHz signal is resonated in winding A, and the voltages are raised. No tuning capacitor is

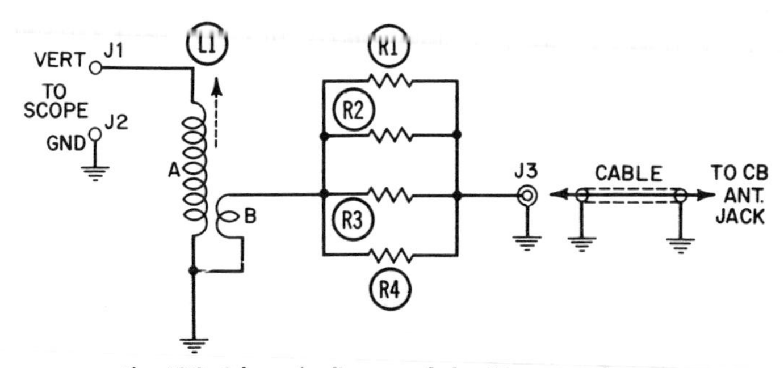

Fig. 13-2. Schematic diagram of the CB scope adapter.

used for winding A, but ample distributed capacity is present for resonating the signal. Output is fed to jack J1, and signal voltage reaches the vertical plate of the scope. The ground connection between scope and adapter is provided by jack J2.

CONSTRUCTION

The major part of assembly is winding coil L1. The large winding (A) consists of seven turns of No. 20 enamel-covered wire. Scrape off the enamel on one end of the wire and solder it to the coil terminal closest to the front panel. The turns may be started in either direction, but be sure to space them out a turn apart (one wire diameter). Clean off the enamel at the end of the winding and solder the wire to the remaining coil terminal. The other winding (B) is simply a length of plastic-insulated hookup wire wound three times over the lower end of the first winding (Fig. 13-3). The turns in this case should touch each other. Shown in Fig. 13-4 is the rear of the completed front panel.

Jacks J1 and J2 are red and black, respectively. J1 runs to the scope plate and must be insulated from the metal panel. Fiber shoulder washers are usually provided with jacks (or binding posts) of this type for insulating purposes. No washer is used on J2 since the jack must make electrical contact with the metal case.

Jack J3 is the input connector that receives rf energy from the CB rig. Use a component here that matches the style connector used on the set. A short length of 52-ohm coaxial cable with male plugs on both ends completes the connections.

OPERATION

After the CB rig is coupled to jack J3 through the coaxial cable, the connections to the scope are made. Use two lengths of hookup wire from J1 and J2. It is important not to use coax between scope and adapter since signal losses will occur. The two wires should be several inches apart.

The wire from J1 connects to the scope's vertical plate, not to the vertical input. The latter connection introduces

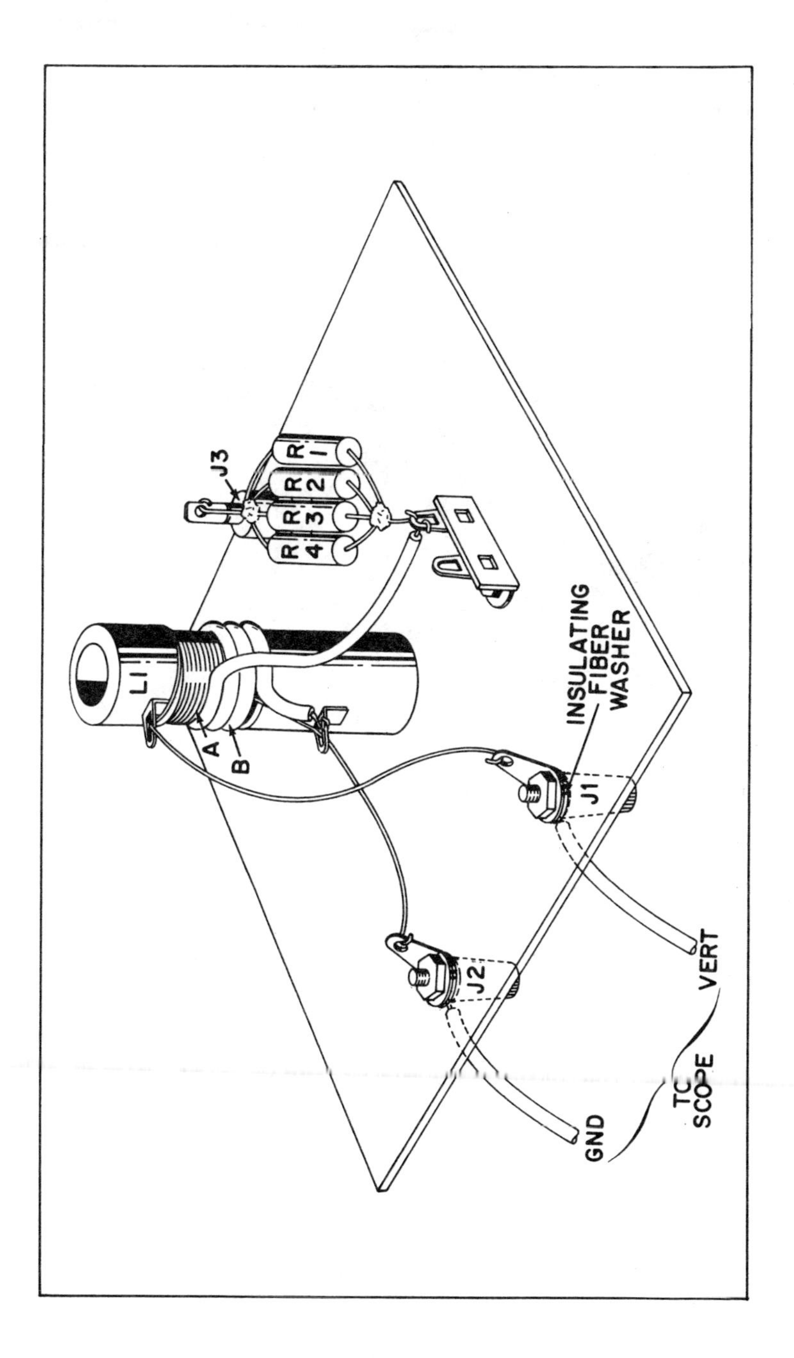

Fig. 13-3. Pictorial diagram of the CB scope adapter.

the rf signal to the vertical amplifiers and no indication will appear on the screen. The ground wire from J2 goes to any ground post on the scope. Once these connections have been made, turn the transmitter on and observe the screen. (The scope Vertical-Gain control should be fully off anytime the adapter is in service.) If the device is operating properly, a band of light appears across the face of the

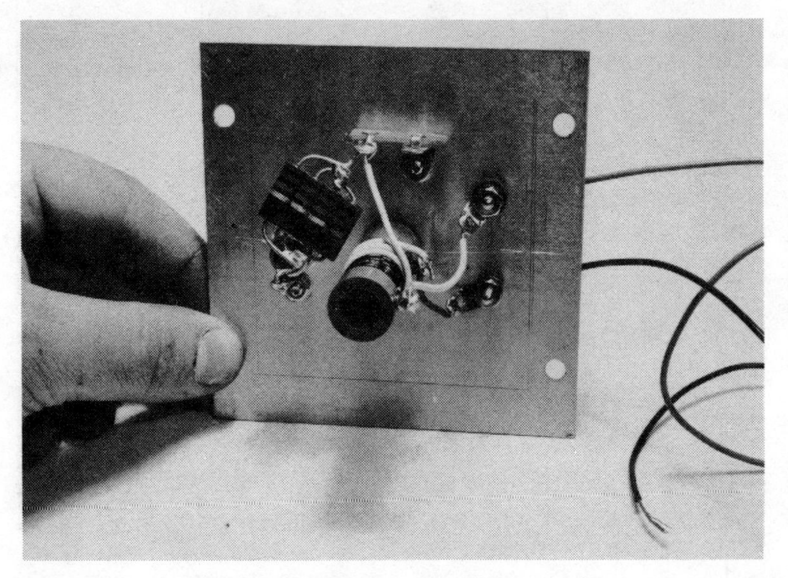

Fig. 13-4. Back side of completed front panel.

screen. Try to make the band as wide as possible by rotating the tuning control, or slug, of coil L1. You should be able to rotate the knob in one direction and make the band narrow and thicken as the coil goes through resonance. If the pattern on the screen will not vary in this fashion, the coil may be adjusted by altering the turns spacing. Begin by setting the coil slug about halfway into the form. Next, squeeze or pull apart the exposed turns on winding A until you can make the pattern on the screen go through a peak. Since your hand will detune the coil, this should be done in a series of trials, removing your hand while observing the screen. When the correct spacing is found, the coil turns are cemented in place with dope. The slug will now permit the proper range of adjustments by the knob on the front panel. A readable scope display on a typical 5-inch screen should be more than an inch high.

Four photos shown in Fig. 13-5 are actual screen traces of a CB signal while using the adapter. They may be interpreted as follows:

In Fig. 13-5A, the rf carrier is shown with no modulation. It is the characteristic envelope trace that occurs when no sound enters the microphone. If any ripples appear at the top or bottom edges of the pattern while the mike is covered with the hand, this may indicate hum or other disturbances are occurring in the transmitter.

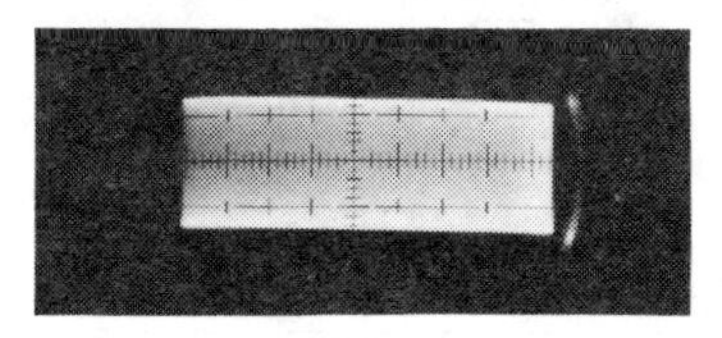

(A) Unmodulated carrier.

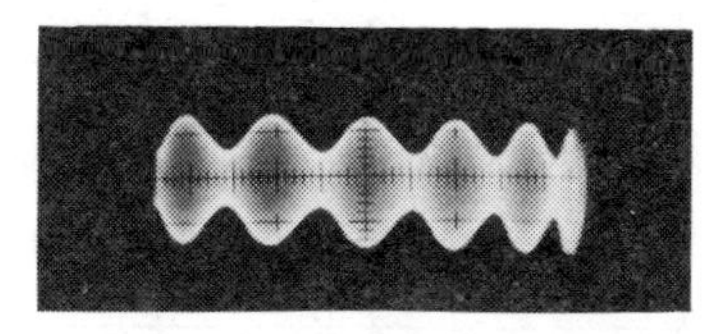

(B) 50% modulated carrier.

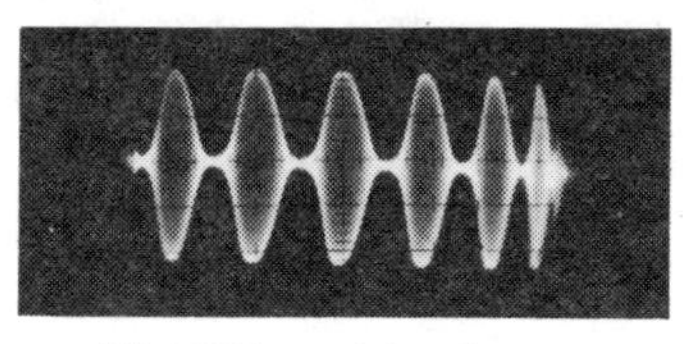

(C) 100% modulated carrier.

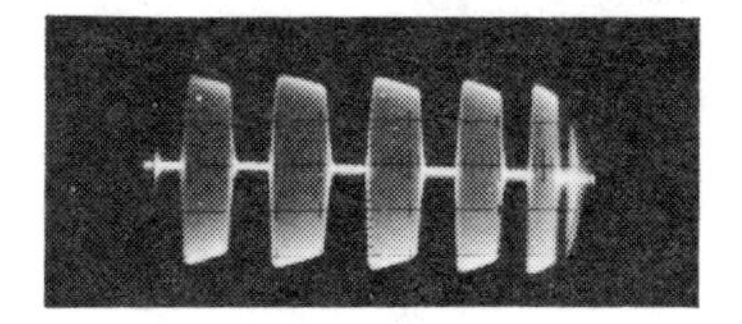

(D) Overmodulated carrier.

Fig. 13-5. Scope traces of percentage of modulation.

Relative power output of the transmitter is indicated by the height of the pattern. If adjustments are made to the rig, the light band will thicken when power increases.

Fig. 13-5B shows a pattern representing 50% modulation. Note how the pattern pinches at the top and bottom. Imagine a zero line running through the horizontal center of the pattern. When the top and bottom of the trace is compressed halfway down to the zero mark, 50% modulation is indicated. Other percentages may be judged in similar fashion.

In Fig. 13-5C, when top and bottom loops compress completely to the zero line, 100% modulation occurs. In Fig. 13-5D, note how the zero line, instead of appearing as a series of short dashes, begins to lengthen. This represents overmodulation, which is illegal in CB operation. The distortion products created during this condition cause excessively broad signals and interference termed "splatter." Note, too, that the curving loops are squared off at top and bottom.

All of these patterns may be generated by speaking at varying distances into the CB microphone. The horizontal frequency (or sweep) control of the scope is set so that the fewest number of loops occur in the modulation. The voice, however, contains many frequencies, and this causes the pattern to shift and blur. One technique is to hum steadily into the mike. A superior system is feeding a constant audio tone of about 400 or 1000 Hz into the CB rig.

PARTS LIST

Item	Description
R1, R2, R3, R4	220-ohm, 1-watt resistors.
L1	Slug-tuned, coil form, ½" dia. (National XR-50 or equiv.) Winding A, 7 turns No. 20 enamel wire, Winding B, 3 turns plastic insulated hookup wire.
J1, J2	Binding posts (J1 requires fiber washer for insulation from chassis).
J3	Jack to match CB output.
Misc	Aluminum case, 4" × 4" × 2"; coax cable with connectors; 1-lug terminal strip; wire.

14

Auxiliary Control Unit

It is often convenient to operate a CB unit when it is several feet from its permanent mounting position. Let's assume that the location of the rig is in a living room or den. If the operator decides to spend several hours in a basement workshop or garage, for example, the rig is temporarily out of service. Incoming calls are missed and the ability to transmit at any time is limited. Controlling the rig from a remote point is possible with the auxiliary control unit (Fig. 14-1). It provides transmit and receive facilities at any point up to 50 or 60 feet from the main unit. The distance is adequate to connect almost any two points in a typical home.

The system works with those CB transceivers with a push-to-talk button on the microphone. It is not suitable for sets with a manual send-receive switch on the front panel.

CIRCUIT DESCRIPTION

The device extends microphone and speaker signals over a pair of cables. A remote speaker and volume control are provided to reproduce the audio signal at a desired level. An arrangement of plugs and packs permits the microphone and speaker to be connected into the CB set in seconds without disturbing normal operation.

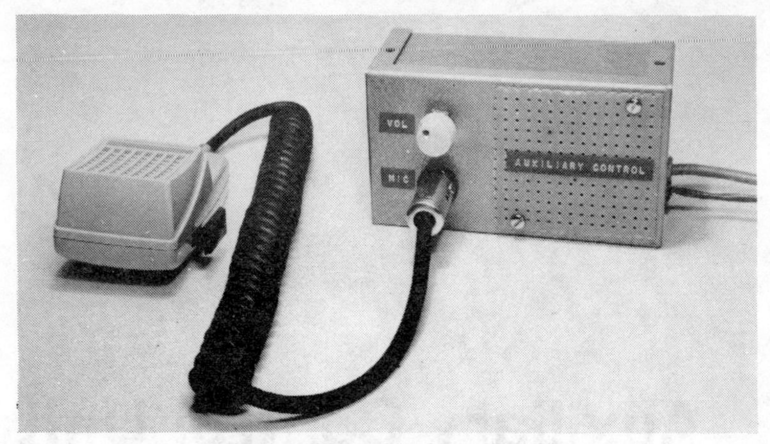

Fig. 14-1. The auxiliary control unit.

A key limitation in the system is the length of the microphone cable. Since most CB sets use ceramic or other high-impedance mikes, there is some susceptibility to hum pickup on the cable. Tests, however, indicate that the suggested cable length works well, and no hum problem exists.

CONSTRUCTION

The size of the aluminum case is governed by the remote speaker (Fig. 14-2) ; a 2½″ speaker is shown, but any standard diameter is suitable. It should be a permanent-magnet type with an impedance of 3.2 ohms. Before the speaker is

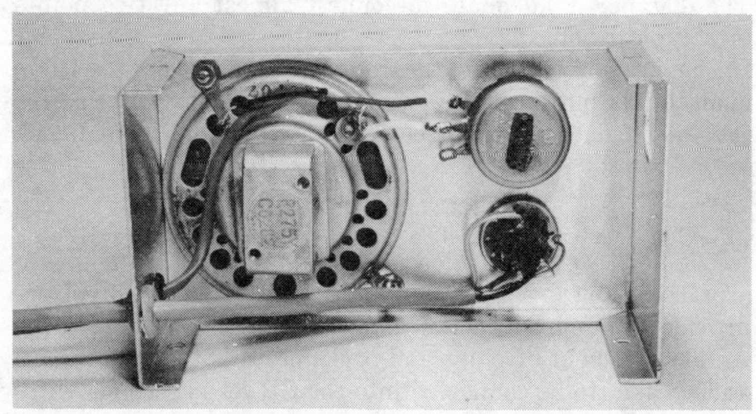

Fig. 14-2. Inside view of completed unit.

mounted, an opening is cut into the front panel to permit passage of sound. A series of small holes, made with a tube socket punch, if one is available, works as well as one large cut-out. Covering the speaker opening is done with a piece of protective material like grill cloth or perforated board. The same screws that hold the speaker in place are used to secure the grill to the panel.

Figs. 14-2 or 14-3 do not show connections required for wiring PL2 and J1. The reason is that a great variety of mike circuits appear in CB sets. Wiring the mike plug and socket, however, is not difficult since they duplicate existing connections in the CB mike cable. The following procedure may serve as a guide.

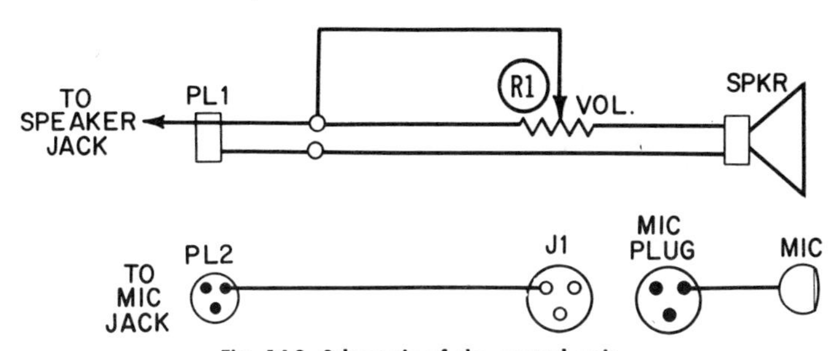

Fig. 14-3. Schematic of the control unit.

PL2 is identical to the plug on the original mike cable. Similarly, the cable connected to PL2 must duplicate the type used in the original mike. Although the microphone may have a coil-type cord, similar cable is available in straight lengths. (For example, in the model, the mike cable has one shielded lead plus two other conductors.) If there is any question of wire size, choose No. 20.

After the desired cable length is selected, connect PL2 to one end in precisely the same manner as found on the original mike plug and cord. The proper terminals may be located by loosening the set screw on the CB mike plug, pulling back the protective cap and viewing the terminal arrangement. The remote, or other end of the cable, runs to J1 on the unit. J1 is the same type of connector used for the mike jack on the CB set. Wire the cable to it, also using the same terminals as in the CB mike jack. In effect the cable is now an extension cord for the microphone.

The second cable carries the speaker signal from the CB set to the auxiliary unit. The wire, consisting of two No. 18 conductors, may be ordinary lamp or zip cord. Install PL1 on the end of the cable that terminates at the CB set.

PL1 is plugged into a jack on the CB set to pick up speaker power. To install the jack, follow the instructions already given in the earlier chapter, Remote Speaker. Inserting PL1 automatically shuts off power to the CB speaker and diverts it into the speaker line.

A final step in construction can be a provision for hanging the CB microphone during remote operation. If the mike has a knoblike projection on its case, an appropriate hole drilled in the cabinet will accept it. The completed auxiliary control unit with the mike can be mounted on a suitable vertical surface by means of wood screws installed through holes drilled in the rear cover of the case.

OPERATION

After the cables are routed and the unit installed at the remote location, the first step for placing it into operation is inserting the plugs. PL2 is placed in the CB-set microphone jack, and PL1 in the speaker jack. The microphone is then taken to the remote position and jacked into J1. This procedure is followed any time auxiliary operation is used.

The controls on the CB set are adjusted in advance for remote operation; the volume control is placed in the normal listening position, power is turned on, squelch is set, and the desired channel is selected. Transmit and receive functions on one channel are now possible at the auxiliary unit. The remote volume control (R1) is used to set audio at proper listening level. If maximum audio is too soft, the volume control on the CB unit should be set at a higher point.

PARTS LIST

Item	Description
R1	100-ohm potentiometer, wirewound.
SP	Speaker, 3.2-ohm.
PL1	Phone plug.
PL2, J1	See text.
Misc	Aluminum case, size optional; cables (see text); rubber grommets.

15

Coaxial Switch

Here is an excellent accessory for anyone who has accumulated more than just a basic CB transceiver and an antenna. The coaxial switch (Fig. 15-1) goes together in less than an hour and is ready to serve in several ways. As illustrated in Fig. 15-2A, the first application is connecting

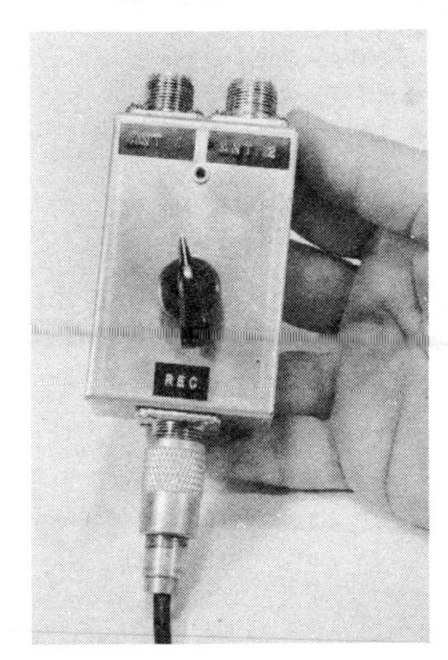

Fig. 15-1. The coaxial switch.

two antennas to one CB set. This is useful when you have both a nondirectional antenna (the usual kind) and a directional type, such as a beam. Instead of loosening and refastening connectors on various cables, the switch does it instantly. This means you can do routine monitoring on the conventional antenna, which picks up signals from any

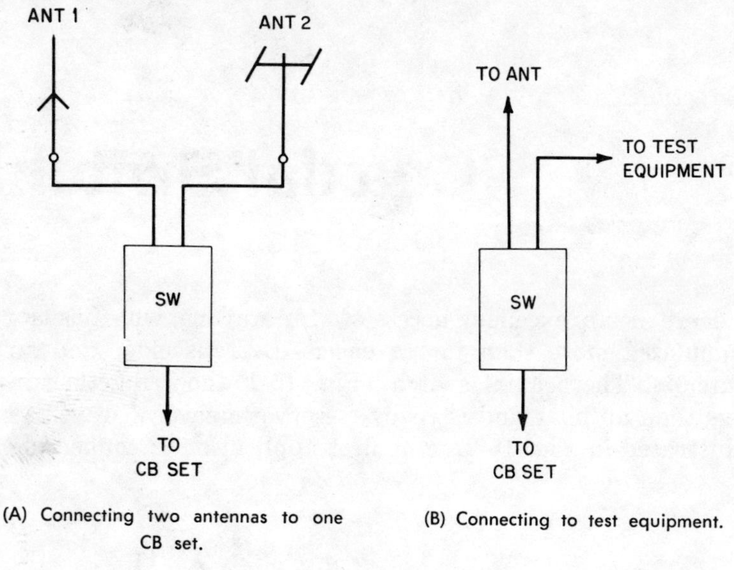

(A) Connecting two antennas to one CB set.

(B) Connecting to test equipment.

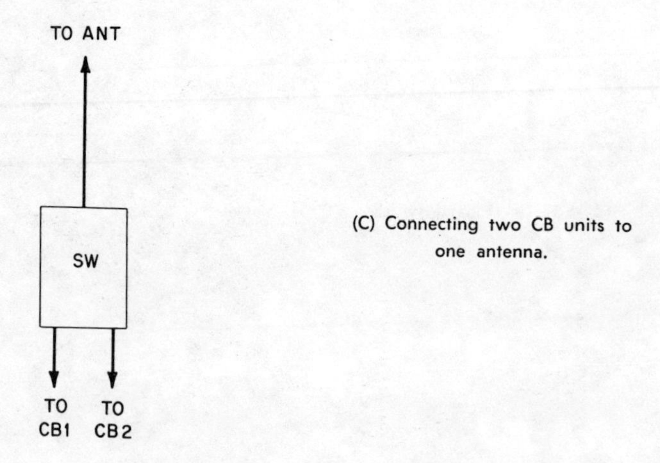

(C) Connecting two CB units to one antenna.

Fig. 15-2. Uses for the coaxial switch.

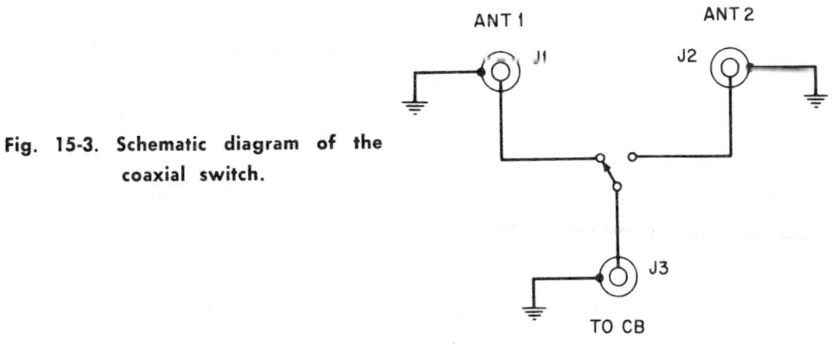

Fig. 15-3. Schematic diagram of the coaxial switch.

direction. When you wish to improve reception and transmission on a signal you hear, use the coaxial switch to make the changeover to a beam. The beam is then swung to favor the desired station.

Another application, as in Fig. 15-2B, is to use one antenna socket for a piece of test equipment, say, an output power meter. If you suspect something is wrong with your signal, switch to the meter for instant verification of power. A new antenna to be tested can also be plugged into that second socket. This permits you to run quick comparisons between old and new antennas, while receiving reports from a distant station.

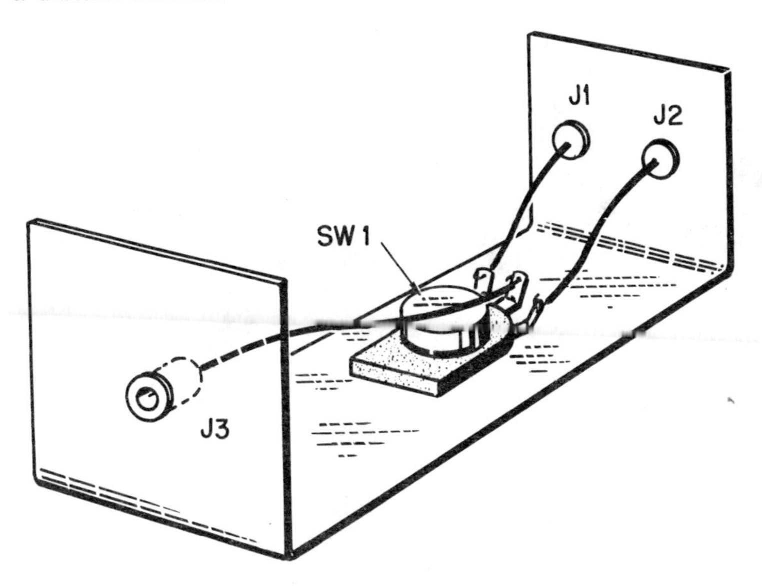

Fig. 15-4. Pictorial of the coaxial switch.

Yet another use for the instrument is shown in Fig. 15-2C. Here the switch is used in reverse—the antenna is now connected to the side with a single socket. At the other end, the two sockets (J1 and J2) are connected to *two* CB sets. This enables you to rapidly connect either of the two sets to the antenna. One set could be a sideband rig, a handie-talkie or some other piece of equipment like a vhf monitor receiver.

CONSTRUCTION AND OPERATION

The schematic of the coaxial switch is given in Fig. 15-3. It is a rotary switch with two positions to divert a signal to one of two connectors. Wiring the device is done by following the pictorial layout given in Fig. 15-4. (The various components and case before construction appear in Fig. 15-5.) In preparing the case, three holes of ⅝-inch diameter

Fig. 15-5. Components before mounting.

are cut for mounting the coaxial sockets. This is considerably simplified if you have a tube-socket punch for this diameter. Otherwise, smaller holes are bored with an electric drill, then widened with a rattail metal file or reamer. The only precaution in wiring the circuit is to use short, direct connections. Any loops or hairpins formed by the wire are apt to cause power losses. Solid copper wire of No. 18 gauge should work well. When installing the wire, avoid running it closer than about a half inch from any surface on the case. To use the completed coaxial switch, simply refer back to the typical applications described earlier.

PARTS LIST

Item	Description
J1, J2, J3	Coaxial connectors.
SW1	Rotary switch, 1-circuit 2-position type (Spdt).
Misc	Aluminum case, 3¼" × 2¼" × 1⅛" approx; No. 18 solid copper wire.

16

Mobile Signal Monitor

When something goes wrong with a base station rig, there is often a piece of test equipment nearby for checking the output signal. In mobile operation, testing is not so convenient. This accessory solves that problem. It tells instantly whether the transmitter is putting out a well-modulated signal. It needs no permanent connection in the CB transceiver and is simply mounted under the dashboard at any location.

The mobile signal monitor uses a flashing lamp to tell whether modulation is satisfactory. Whenever you speak into the mike, the lamp glows with each syllable. This is even more informative than a field-strength meter or other output indicator which measures only the presence of an rf carrier. For this monitor to flash, *both* carrier and audio modulation must be present. Another bonus is the high visibility of the lamp. You don't have to look directly at it, as you would a meter, to check the signal. This is an advantage at night since lamp flashes are easy to see in the dark.

The instrument is extremely sensitive, but not difficult to build. The reason is that much of the construction centers around a factory prefabricated module which contains a complete audio amplifier (Fig. 16-1). This reduces construction to some uncomplicated input and output circuitry. These audio amplifiers are widely distributed and are rea-

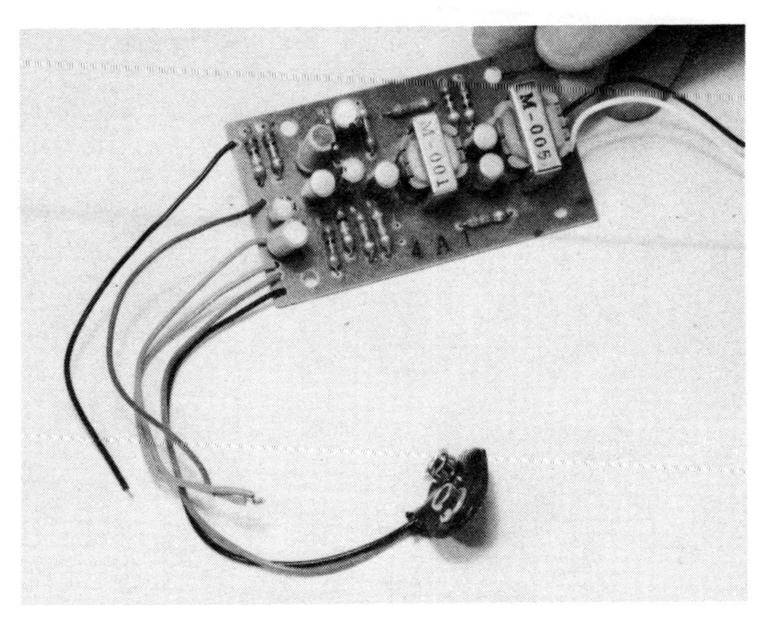

Fig. 16-1. Audio amplifier module.

sonable in cost. The one shown here produces 100 milliwatts of audio output with 1 millivolt input and is normally used as the basis of a simple phono amplifier or intercom. Shown in Fig. 16-2 is the lead layout of a typical amplifier of this type (shown in catalogs as a "100 milliwatt 4-transistor amplifier"). Along the left side are green and black leads

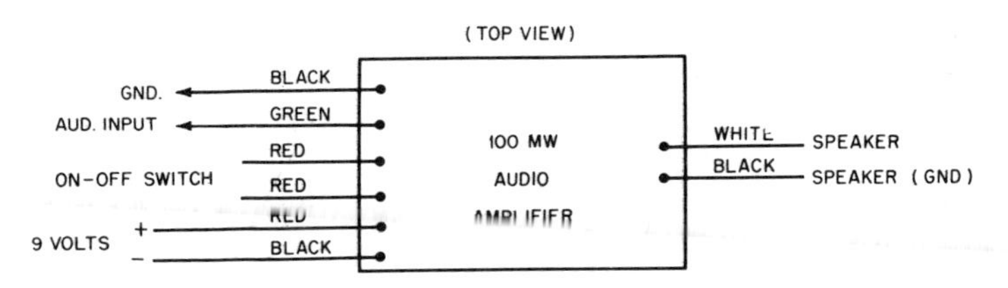

Fig. 16-2. Original board connections.

for accepting the audio input. There is a pair of red leads which connect to an On-Off switch for controlling power. Red and black leads go to a 9-volt battery source, while white and black leads connect to a speaker output.

Our project uses all those leads, but with a different input and output arrangement. The audio input is replaced by the demodulated radio signal from the transceiver. The speaker is supplanted by the flashing lamp.

CIRCUIT DESCRIPTION

The schematic in Fig. 16-3 shows how the circuit functions. The antenna senses the CB signal and applies it to the 27-MHz circuit formed by capacitor C1 and coil L1. The rf signal is converted to an audio voltage by diode D1. The

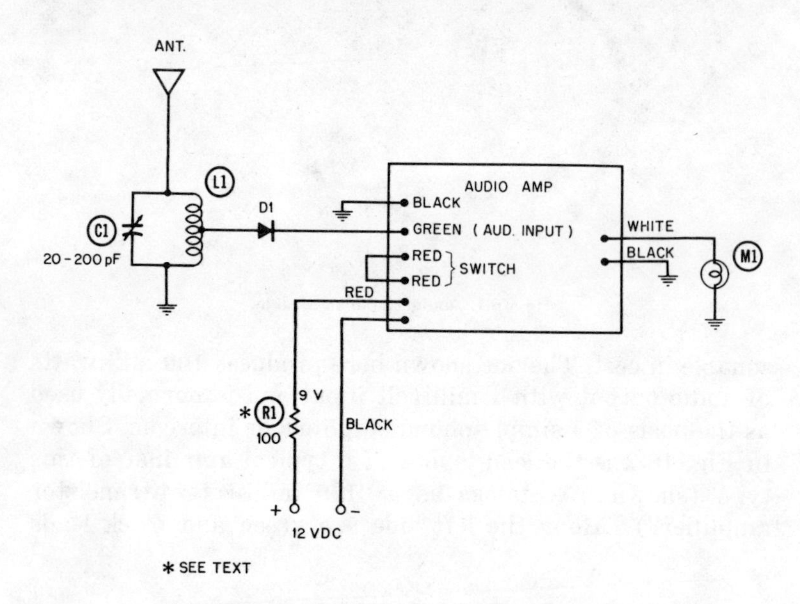

Fig. 16-3. Schematic diagram of mobile signal monitor.

complete schematic of the audio amplifier is not shown because it is a prefabricated module which needs no construction. Only the appropriate connections are given. After the signal strength is built up, it is applied to M1, the flashing indicator lamp. The lamp absorbs audio energy as would a speaker, but the energy is converted to light, rather than sound.

Note the use of an optional resistor, R1, in the positive side of the power supply. If the device will be used on a 9-volt battery, no resistor is needed. This might apply if you want to use the instrument as a base station monitor. In

Fig. 16-4. Mounting the tuning capacitor and lamp.

mobile work, however, it is handier to power the device with the car's electrical system. Since that supply is about 14 Vdc, the 100-ohm resistor is inserted in the line to keep the supply voltage down to safe levels.

CONSTRUCTION

Begin assembly by fitting the amplifier module into the plastic box. The original model required some notching and fitting of the module board to make it lie within the narrow confines of the cabinet. This can be done if you are careful not to damage the foil conductors printed on the board; or, you can use a larger case and avoid any cutting. Keep the case small, however, because space under the dashboard may not be very generous in some cars.

Tuning capacitor C1 takes some special mounting techniques (Fig. 16-4). Two saw cuts made with a hacksaw blade provide vertical slots in the plastic case into which you can slide the two solder lugs on the capacitor. Before putting C1 in place drill a hole between the two saw cuts to receive the tuning screw of the capacitor. Also, break off two tiny metal tabs at the underside of the capacitor so it fits snugly against the case. Note that the ground lug of the capacitor should lie in the direction of the lamp (M1). You can identify the lug by looking at the screwhead of the capacitor; the ground side goes to the outermost plate. Solder wires to the tip and shell of the lamp (using very little heat) and slide the bulb into a hole cut into the case, as shown in Fig. 16-4. Retain the lamp in the hole by bedding it in some glue. The wire from the shell of the lamp connects to the ground lug of the tuning capacitor. Complete work on the case by drilling a hole at the rear for the two power leads (red and black), and drill a hole next to the hot lug of the tuning capacitor to allow the antenna wire to pass to the outside.

The amplifier module is prepared by soldering together the two red leads which are intended to connect to an On-Off switch. (Your module might use a slightly different arrangement or color code, but it shouldn't be difficult to find the equivalent connections.) Those switch leads are not used if the monitor is set up for mobile operation, as we will see later on. If you plan to use the instrument at a base station and powered by a dry-cell battery, use the red leads with a switch, as they were intended. Slip the module into the case after the lamp and tuning capacitor are in place.

The coil is wound with ordinary No. 20 plastic-coated hookup wire turned on a half-inch diameter form for five turns. Find the center of the coil and slice away a bit of insulation to accept a lead from the diode (D1). Before connecting the diode, tin the bare spot on the coil with a bit of solder and do the same for the diode lead. Spread the coil turns so you can solder that center turn without melting the plastic on adjacent turns (Fig. 16-5). Note that the bar side of the diode (the cathode) is *away* from the tuning coil. Connect that side to the green, or audio input, lead to the module and tape the connection to avoid a short circuit when the metal cover is installed on the plastic cabinet. A solder lug runs from the ground lead of the tubing capacitor to the threaded screw hole of the case (Fig. 16-6).

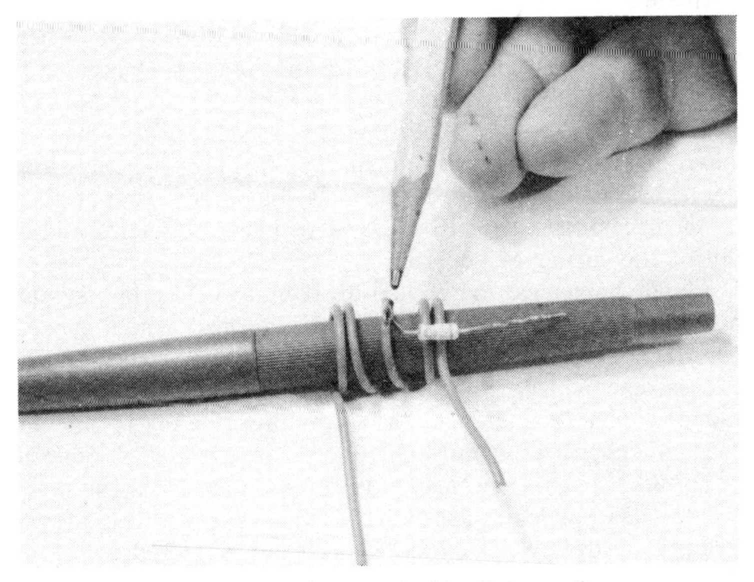

Fig. 16-5. Spread turns and solder diode to coil.

This automatically grounds the metal cover when it is installed and makes the instrument less susceptible to hand capacity while tuning. The completed monitor is shown in Fig. 16-7.

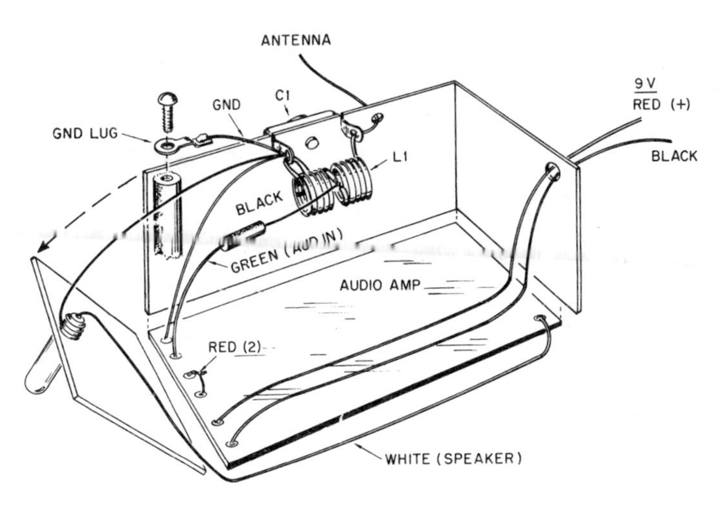

Fig. 16-6. Pictorial diagram.

Installation of the instrument is done with a No. 6 machine screw mounted on the top cover (Fig. 16-8). There are usually some holes already under the dash to receive it. The location of the monitor is not critical and actual tests proved it sufficiently sensitive for good results when located anywhere under the dash. Try to place it somewhere near your field of view, if possible, but direct line-of-sight is not necessary (Fig. 16-9). You will see the flashing lamp out of the corner of your eye.

The dc power can be picked up from the electrical system of the car. One convenient place is the lead which powers the CB set. It is assumed the rig is powered from a 12-volt "accessory" source; that is, turning off the ignition key removes power from the set. Also, turning the key to the "accessory" position permits CB operation without the engine running. If this is the case, wire the red lead from the monitor to that hot lead. Ground the black wire, unless a good ground is already provided by the No. 6 mounting screw on the top cover.

By picking up 12-volt power from an accessory source you eliminate any On-Off switch since the monitor is on

Fig. 16-7. Completed monitor.

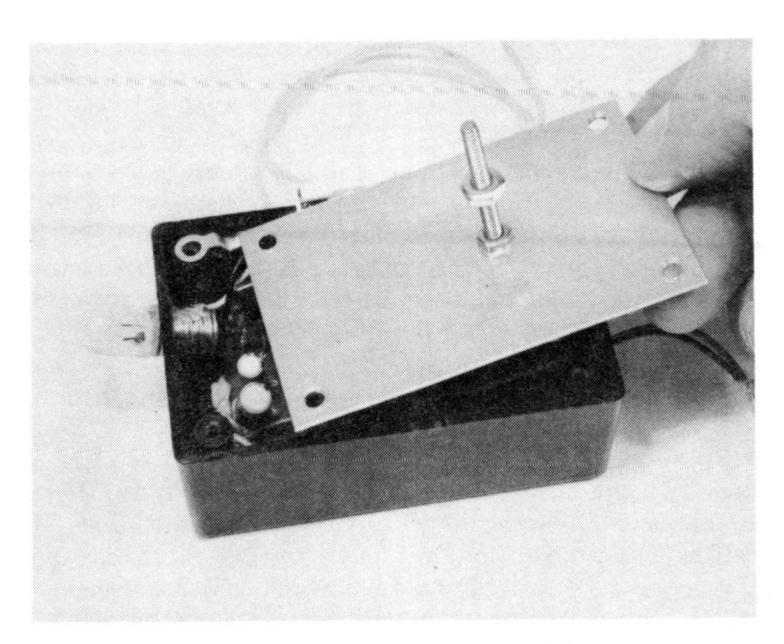

Fig. 16-8. Attach a No. 6 machine screw through hole in cover.

Fig. 16-9. Monitor installed in car.

when the key is turned in the ignition. It doesn't matter if the circuit is on continuously at this time because it draws only 10 milliamperes when no CB signal is being transmitted. Current rises to about 50 mA at full signal (and the lamp is flashing). Install the dropping resistor—a 100-ohm, ½-watt unit—in the red power lead to drop car voltage to the 9 volts required by the module. (If you want to use a 9-volt battery, with an On-Off switch, to power the monitor in module operation, eliminate the dropping resistor.)

OPERATION

To adjust the completed instrument, turn on the CB set and transmit in normal fashion. As you speak, adjust the tuning screw of capacitor C1. When you reach the correct setting, the lamp should flash in step with your voice. Tape the antenna behind the unit, but don't run it all along a metal surface or sensitivity and tuning might be affected. For best pickup you can either coil the antenna wire or cut it shorter depending on how it performs in your particular car.

PARTS LIST

Item	Description
C1	20- to 200 pF mica compression trimmer lapacitor.
L1	Coil, 5 turns No. 20 hookup wire, plastic-insulated, ½-inch diameter.
D1	Diode, 1N60.
M1	2-volt pilot lamp (No. 48 or equivalent).
R1	100-ohm resistor, ½-watt (see text).
Aud amp	Audio amplifier module, 4-transistor, 100-milliwatt output.
Misc	Plastic case, 1⅛" × 3¼" × 2⅛" (Archer 270-230 or equivalent); solder lug.

17

On-The-Air Sign

This on-the-air sign (Fig. 17-1) illuminates whenever the mike button is pressed. Unlike other accessories which do the same thing, this one requires no connection to the CB set. You do not have to locate or identify wires inside the CB transceiver to make the hookup. The device samples a bit of rf energy in the vicinity of the transmitter, amplifies it, and drives a pair of indicator lamps. The pickup is through a short antenna lead wrapped around the coaxial lead emerging from the CB transceiver.

The instrument does more than announce that you are on the air. In responding to an air signal, it also gives direct proof you are actually transmitting. (Conventional on-the-air signs can illuminate even if there's no transmitter output.) Thus, you have a running indication that the transmitter is operating properly.

CIRCUIT DESCRIPTION

As shown in Fig. 17-2 the antenna (ANT) picks up the 27-MHz signal and applies it to a resonant circuit formed by C1 and L1. Capacitor C1 is adjustable for final tuning after the project is completed. Next, the signal is applied

Fig. 17-1. Completed on-the-air sign.

to diode D1 to convert it to steady dc. The rectification process is aided by .001-μF capacitor C2.

The signal developed by the diode is dc negative and is impressed on the base of X1, the first amplifier. The signal is amplified by the tree transistor stages, X1 through X3 until it reaches the pair of indicator lamps, M1 and M2. Two lamps were selected to spread the light behind the "on-the-air" words on the front panel. An advantage of this circuit is low current drain unless the air signal is actually present and driving the lamps.

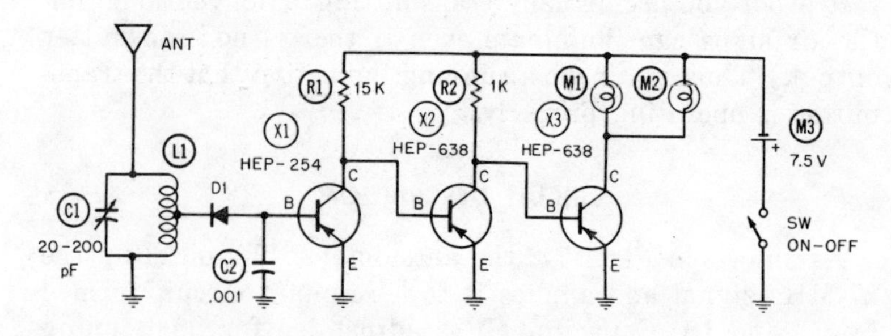

Fig. 17-2. Schematic diagram.

CONSTRUCTION

Most parts are assembled on a small piece of perforated plastic board (Fig. 17-3) which measures 1½″ × 2½″. Enlarge several holes in the board to admit both terminals and mounting tabs of tuning capacitor C1. Bend the tabs to retain the capacitor in place. The two lamps are pressed through holes in the board (Fig. 17-4) and held there by pressure and their leads.

The coil (L1) is made by winding five turns of No. 18 enamel wire around a dowel or other circular form that's ½-inch in diameter. The dowel is removed and the two ends of the coil are scraped free of enamel and soldered to their respective points in the circuit (Fig. 17-5). Also scrape off the enamel at a point two turns from the end of the coil which lies nearest the board. Tin the bare copper there with an iron to make it easy to solder one lead of the diode (D1) to the coil tap. Connect the banded end of the diode toward the coil.

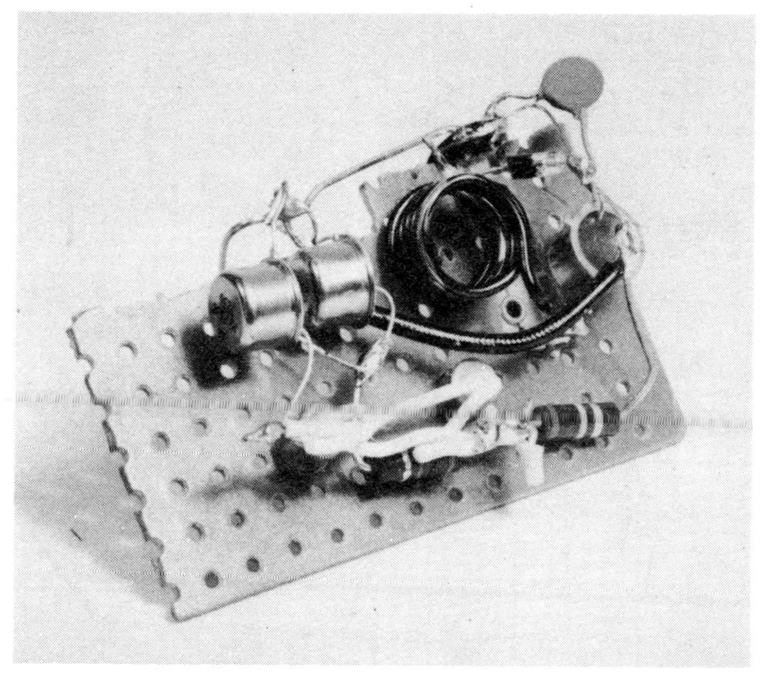

Fig. 17-3. Parts are assembled on perforated plastic board.

Fig. 17-4. Lamps mounted on board.

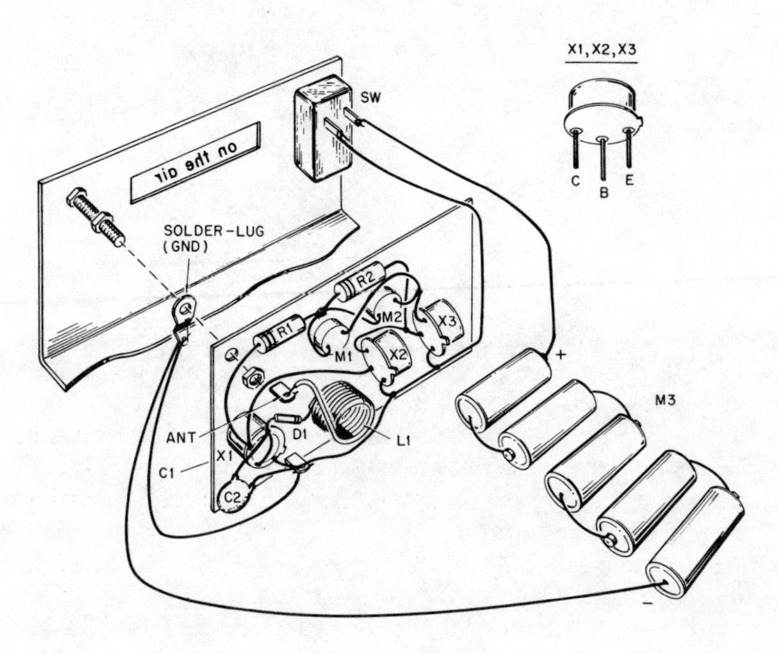

Fig. 17-5. Pictorial diagram.

A precaution about the transistors is to select the correct terminals—emitter, base and collector—as given in the pictorial wiring guide (Fig. 17-5). Also avoid holding the iron too long on transistor leads to prevent possible damage.

Note that a solder lug connects to the circuit ground (the positive side). This lug will contact a ¾-inch long screw (No. 6 size) which holds the board to the metal cabinet. A wire from the lug goes to the ground side of C1, the tuning capacitor. That capacitor is less critical to tune if it is

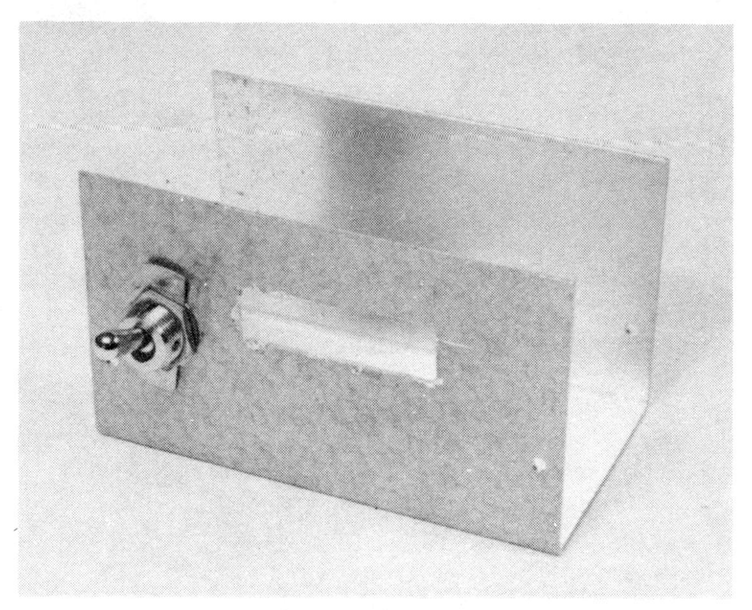

Fig. 17-6. Cut out a window in metal case.

mounted in a certain position with respect to its terminals. Observe the tuning screw on the capacitor and note that it compresses a flat metal plate. That plate should run toward the lug of the capacitor on the ground side of the circuit. Reverse it in its mounting, if necessary, to place the tuning screw at ground potential. After the circuit board is wired, temporarily lay it aside.

Prepare the metal case next. Cut the window (Fig. 17-6) for the on-the-air lettering to about 1½″ × ½″. The exact location is determined by holding the circuit board inside the case and noting how the two lamps are positioned. The

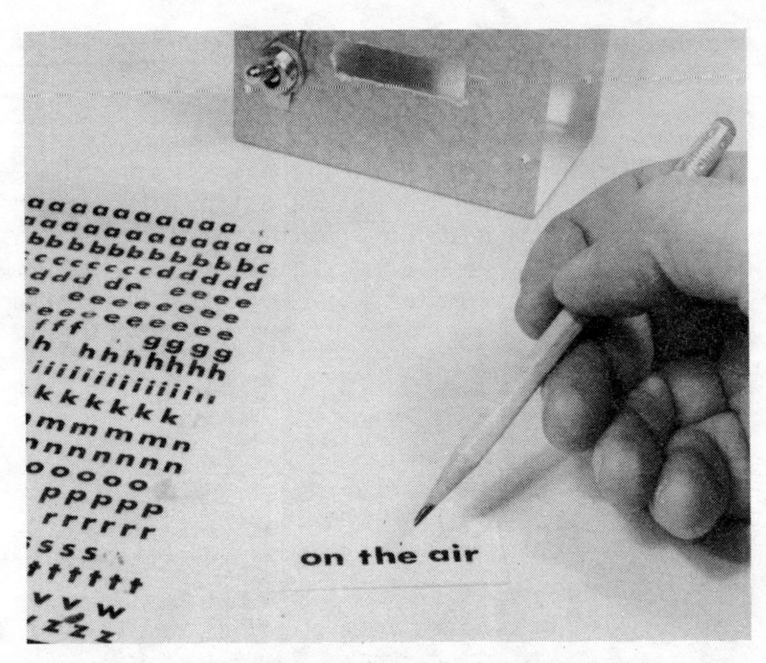

Fig. 17-7. Preparing the works.

Fig. 17-8. Installing the batteries.

material used to support the letters must be translucent to permit light from the lamps to shine through. Very thin white plastic served in the original model, but other material, like white paper should also work. The words are made with "rub-on" letters (Fig. 17-7).

The batteries are five penlite cells wired in series for a total of 7.5 volts. No battery holder is used; simply solder leads from positive to negative terminals and tape the cells together (Fig. 17-8). Cut a hole for the On-Off switch and fasten it. Next, temporarily mount the circuit board with

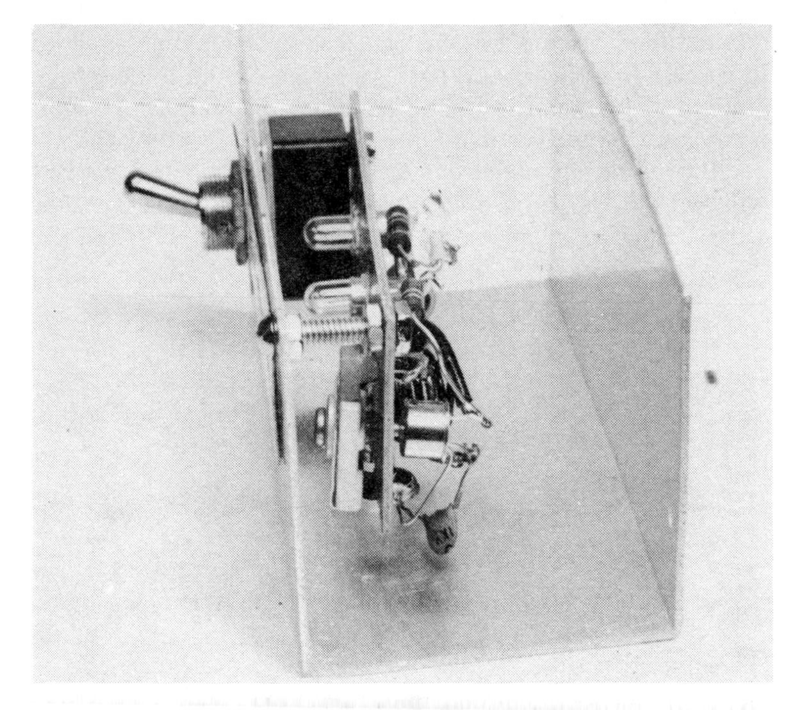

Fig. 17-9. Attaching the circuit board to the case.

the machine screw inserted through the front panel. Fasten that screw with one nut tightened at the rear of the panel. Thread a second nut on to the screw to form a platform which supports the circuit board (Fig. 17-9). The remaining support for the board is at the On-Off switch; it also serves as a platform. If your On-Off switch is not physically similar to the one shown here, simply use a second machine screw to hold the other side of the board. After the board is in place, mark the inside of the front panel directly opposite the capacitor tuning screw. Remove the board and cut a hole at that mark so you can insert a screwdriver later for adjustments. A hole is made at the rear for an antenna wire about 18 inches long.

OPERATION

Place the completed instrument atop, or next to, the CB transceiver. Take the antenna wire and wrap it about a half-dozen turns around the coaxial transmission line plugged into the CB set (Fig. 17-10). As you hold the mike button down and transmit, adjust the tuning capacitor. At some point the lights should illuminate. Don't unscrew the capacitor more than about two or three full turns or the screw will fall out. If you see no light, try adjusting the antenna wire. You may change the number of turns, or slide the

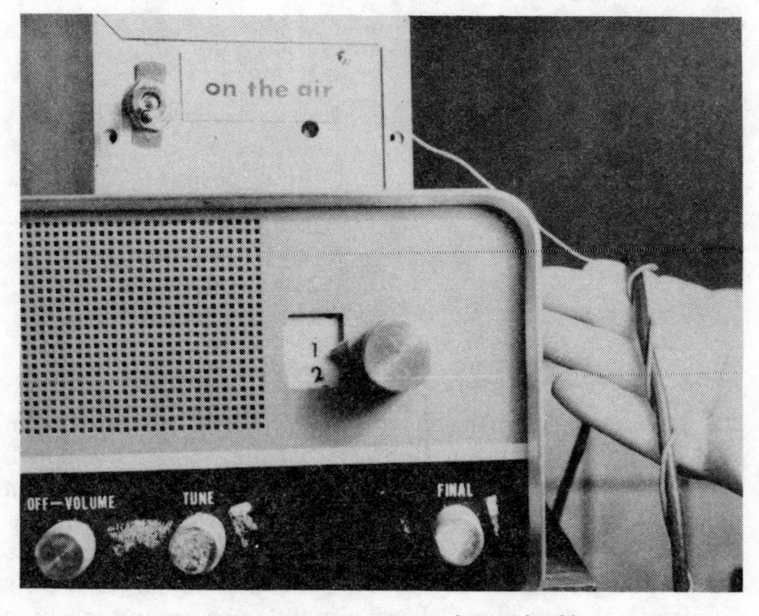

Fig. 17-10. Loop antenna around coaxial cable.

antenna back and forth along the coaxial cable to find a better pickup point. Each time any adjustment is made, retune the tuning capacitor. When the lights go on each time you key the mike, the instrument is operating properly.

It may seem incorrect that a coaxial cable can emit rf energy. Since the cable is shielded, it is usually believed to completely contain the signal. This would be true if the standing-wave ratio of the line were perfect. No system is that exact and the coaxial shield allows a tiny amount of rf energy to escape. That's all it takes to operate this instrument.

PARTS LIST

Items	Description
C1	Mica compression trimmer capacitor, 20-200 pF.
C2	.001-μF disc capacitor.
R1	15K resistor, 1/2 watt.
R2	1K resistor, 1/2 watt.
X1	Pnp transistor (HEP-254 or equivalent).
X2, X3	Pnp transistors (HEP-638 or equivalent).
M1, M2	6V subminiature lamp.
M3	7.5 Vdc battery (5AA penlite cells in series).
L1	Coil, 5 turns, No. 18 enameled wire, 1/2-in. diameter.
SW	Spst switch.
D1	Diode, 1N34 or equivalent.
Misc	Perforated board 1 1/2" × 2 1/2"; minibox approximately 2 1/4" × 2 1/4" × 4".

18

Antenna Tuner

This accessory (Fig. 18-1) is permanently inserted in the coaxial transmission line to correct any mismatch between the line and antenna. With the aid of an SWR meter, the tuner is adjusted until the standing wave ratio is brought to the lowest possible value. This assures that maximum signal is transmitted and received by the antenna system. Although the tuner can be used for base station antennas, it finds greatest application in the trunk of an automobile where it matches the feedline to the mobile antenna.

CIRCUIT DESCRIPTION

With a perfect match between transmitter, coaxial line, and antenna, power loss would be nearly zero. In actual practice, however, mounting an antenna on the car often drops the rating (or impedance) from an ideal 50 ohms to 30 ohms or even less. The result is a downward mismatch which has a short-circuiting effect on the signal. If, on the other hand, the antenna presents more than 50 ohms to the line because of nearby metal objects affecting its operation, part of the energy is similarly lost. In both cases, SWR, or standing wave ratio, increases. The antenna tuner can compensate for either condition.

The specific components are shown in the schematic (Fig. 18-2). The signal from the transmitter encounters

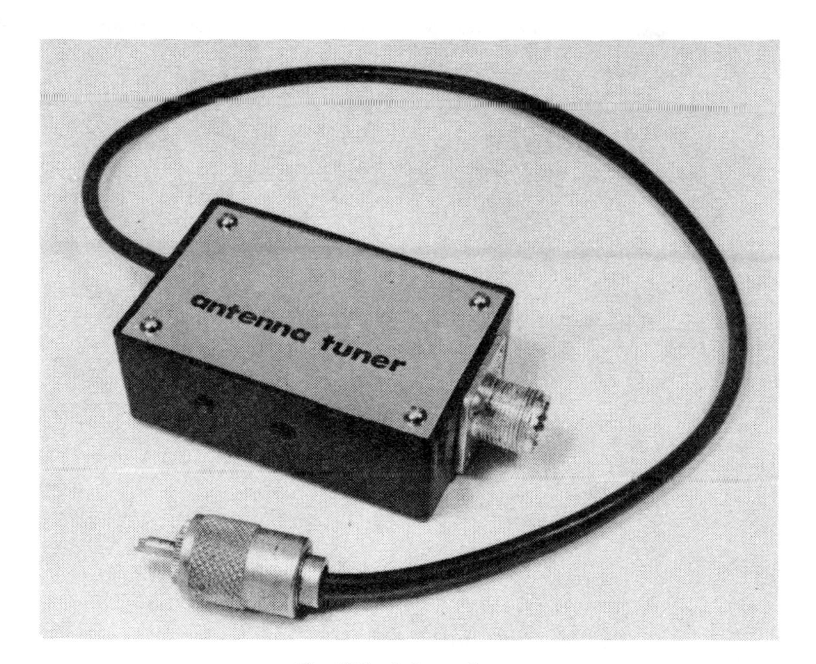

Fig. 18-1. Antenna tuner.

a parallel-tuned circuit consisting of tuning capacitor C1 and coil L1. Next is a series-tuned circuit formed by tuning capacitor C2 and coil L2. The coils and capacitors have identical values to tune to 27 MHz, but are connected in series and parallel to handle a wide range of impedance mismatches.

CONSTRUCTION

The tuner is built in a plastic case with a metal cover. Mount coaxial socket SO-1 after cutting a 5/8″ diameter hole

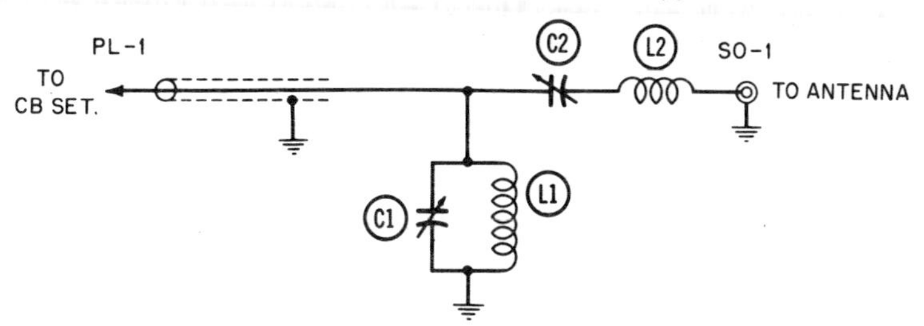

Fig. 18-2. Schematic diagram.

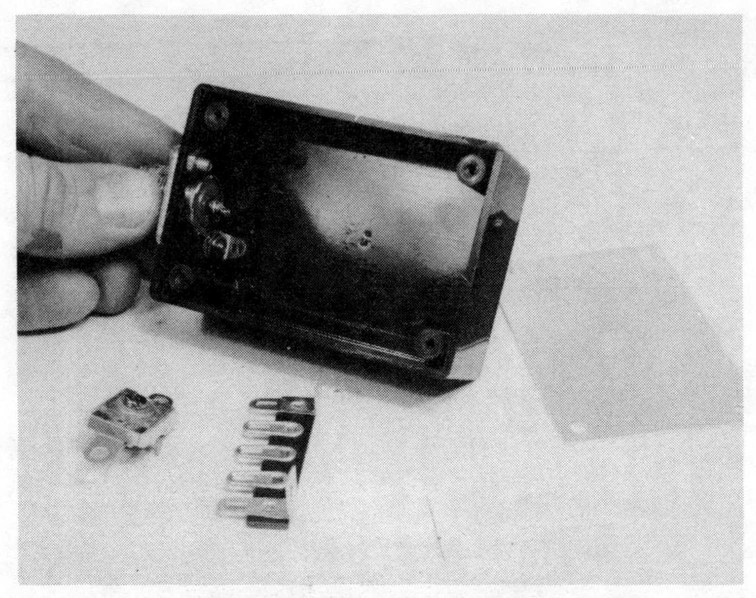

Fig. 18-3. Mount coaxial socket in hole cut in case.

(Fig. 18-3). Next, the five-lug terminal strip is fastened to the bottom of the case. Since some connections are critical, the lugs of the strip are shown numbered from 1 to 5

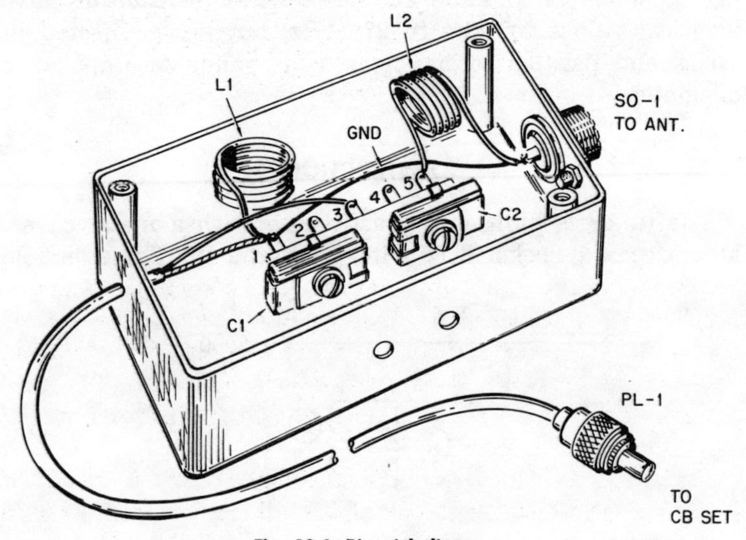

Fig. 18-4. Pictorial diagram.

in the pictorial guide (Fig. 18-4) and wired as follows. (The strip is the type with end lugs which also serve as mounting feet.) The foot of Lug 1 is held to the case with a No. 6 screw and nut. The lower hole in Lug 1 also takes the shield of the coaxial cable. Note that Lug 5 also has a mounting foot. Do not bolt this to the case with a screw. The foot must be insulated and not be permitted to touch anything outside the cases through a mounting screw (Fig. 18-5). One way to secure Lug 5 is with a dab of cement.

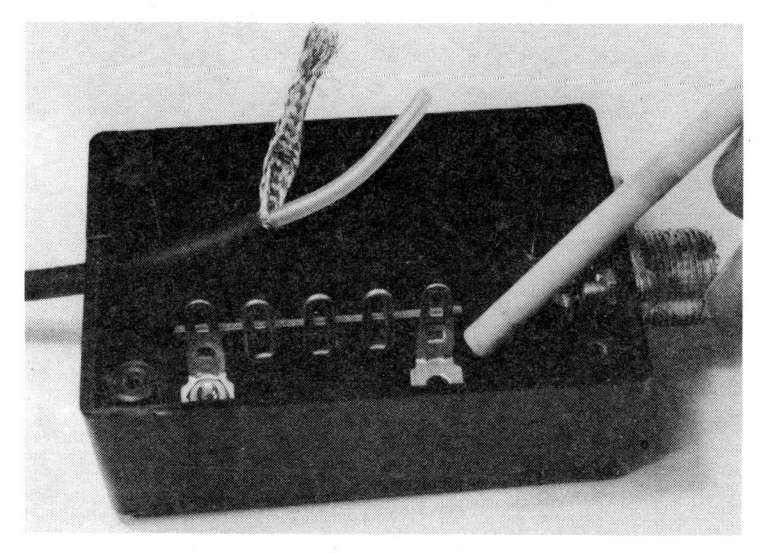

Fig. 18-5. Mount terminal strip.

After these steps, clip open the top loops of Lugs 1, 3 and 5. This makes it easier to install the mounting tabs on the two trimmer capacitors (Fig. 18-6). To prevent interference, bend Lugs 2 and 4 down toward the case, out of the way. Note that the center (hot) lead from the coaxial cable is soldered to Lug 3 on the terminal strip. To avoid interference with other components, solder this wire to the lower opening on Lug 3.

The coils are identical and fashioned from 5 turns of No. 20 plastic-insulated hookup wire. Turn the wire around a ½-inch form, slide off the coil and solder it to the appro-

priate lugs. Note the orientation of the coils in Fig. 18-7; L1 has an open end facing one way, L2 faces in another direction. This prevents coupling between them and poor circuit action.

Fig. 18-6. Mount the trimmer capacitors.

OPERATION

Install the antenna tuner as close to the base of the antenna as possible. An SWR meter, essential for adjusting the antenna tuner, is inserted in the line between the CB set and the antenna tuner. Although we have shown the cable end (PL-1) of the tuner pointing toward the CB set and the socket end toward the antenna, it is not necessary to connect the antenna tuner in this direction. It is reversible in the circuit with no effect. This allows you to use the coaxial plug and socket for easiest mechanical connection into your system.

Once the tuner and SWR meter are in place, turn on the transmitter and adjust the two tuning capacitors through the access holes in the tuner case (Fig. 18-8). Watch the

Fig. 18-7. Note orientation of the coils.

Fig. 18-8. Adjust tuning capacitors.

SWR meter as you tune each screw. There will be much interaction between them, so keep adjusting back and forth until you see lowest SWR on your meter. Once it is found, remove the SWR meter and leave the tuner in the line. In tests with the unit shown here, the tuner could overcome a considerable range of mismatch. In one trial, the 50-ohm transmission line was connected to a load of 120 ohms for more than a 2-to-1 upward mismatch. The tuner easily brought SWR down to a negligible 1.1 to 1. When the 50-ohm line was connected to a load of 22 ohms, a downward mismatch of more than 2-to-1, the tuner eliminated the mismatch to the point where SWR could not be seen on the SWR meter.

PARTS LIST

Items	Description
C1, C2	Mica compression trimmer capacitors, 20-200 pF.
L1, L2	Coil, 5 turns No. 20 plastic-insulated hookup wire, 1/2" dia.
PL-1	Coaxial connector, Pl-259.
SO-1	Coaxial connector, SO-259.
Misc	Plastic case, 1 1/8" × 3 1/4" × 2 1/8" (Archer 270-230 or equivalent); 2 feet of RG/58U coaxial cable.